Genetics, Genomics and Biotechnology of Plant Cytoplasmic Organelles

Genetics, Genomics and Biotechnology of Plant Cytoplasmic Organelles

Editors

Nunzia Scotti
Rachele Tamburino

MDPI • Basel • Beijing • Wuhan • Barcelona • Belgrade • Manchester • Tokyo • Cluj • Tianjin

Editors
Nunzia Scotti
Institute of Biosciences and
BioResources
Italy

Rachele Tamburino
Institute of Biosciences and
BioResources
Italy

Editorial Office
MDPI
St. Alban-Anlage 66
4052 Basel, Switzerland

This is a reprint of articles from the Special Issue published online in the open access journal *Plants* (ISSN 2223-7747) (available at: http://www.mdpi.com).

For citation purposes, cite each article independently as indicated on the article page online and as indicated below:

LastName, A.A.; LastName, B.B.; LastName, C.C. Article Title. *Journal Name* **Year**, *Volume Number*, Page Range.

ISBN 978-3-0365-2648-5 (Hbk)
ISBN 978-3-0365-2649-2 (PDF)

Contents

About the Editors

Nunzia Scotti is a Senior Researcher at the Institute of Biosciences and BioResources, National Research Council of Italy. She graduated in Biological Sciences and completed her PhD in Molecular Systematics at the University of Naples Federico II. She has been visiting scientist at the CNRS Institute of Plant Molecular Biology, Strasbourg, France and at the Max Plant Institute of Molecular Plant Physiology, Potsdam, Germany. Dr. Scotti has published in several peer-reviewed journals, book chapters and conference papers. Her research interests are mainly focused on biology and biotechnology of cytoplasmic organelles, mostly related to the analysis of the organelle genome variability, to the production of recombinant proteins and to study fundamental biological aspects of plant organelles.

Rachele Tamburino is a Researcher at the Institute of Biosciences and BioResources, National Research Council of Italy. She graduated in Food and Industrial Biotechnology and received her PhD from Università della Campania "L. Vanvitelli". Her research deals with biochemistry and molecular biology applied to cytoplasmic organelles to investigate fundamental biological aspects of chloroplasts and mitochondria. She has published in several international peer-reviewed journals and conference papers.

Preface to "Genetics, Genomics and Biotechnology of Plant Cytoplasmic Organelles"

Chloroplasts and mitochondria are essential plant organelles involved in many fundamental processes (photosynthesis, energy production, metabolism, cell homeostasis, etc.). They are semi-autonomous organelles; hence, their function is dependent upon cross-talk between nuclear and organelle genetic systems. The Special Issue Book entitled "Genetics, Genomics and Biotechnology of Plant Cytoplasmic Organelles" provides new advances in the sequencing of both mitochondria and chloroplasts' genomes using Next-Generation Sequencing technology in plant species and algae including important crop and tree species, in vitro culture protocol, and identification of a core module of genes involved in plastid development. In particular, the published studies focus on the description of adaptive evolution, elucidate mitochondrial mRNA processing, highlight the effect of domestication process on plastome variability and report the development of molecular markers. A meta-analysis of recently published genome-wide expression studies allowed the identification of novel nuclear genes, involved in the complex and still unrevealed mechanisms at the basis of communication between chloroplast and nucleus (retrograde signalling) during plastid development (biogenic control). Finally, an optimized regeneration protocol useful in plastid transformation of recalcitrant species, such as sugarcane, has been reported. We would like to thank to all the authors and their contributions as well as the reviewers for their constructive suggestions. Further, the Plants MDPI team and, particularly, Hinata Fang deserve our thanks for their support.

The editors declare no conflict of interest.

Nunzia Scotti, Rachele Tamburino
Editors

Article

Mitochondrial Genome of *Fagopyrum esculentum* and the Genetic Diversity of Extranuclear Genomes in Buckwheat

Maria D. Logacheva [1,2,*], Mikhail I. Schelkunov [1,2], Aleksey N. Fesenko [3], Artem S. Kasianov [1] and Aleksey A. Penin [1]

[1] Institute for Information Transmission Problems of the Russian Academy of Sciences, 127051 Moscow, Russia; m.shchelkunov@skoltech.ru (M.I.S.); artem.kasianov@gmail.com (A.S.K.); alekseypenin@gmail.com (A.A.P.)
[2] Skolkovo Institute of Science and Technology, 143026 Moscow, Russia
[3] Federal Scientific Center of Legumes and Groat Crops, 302502 Orel, Russia; fesenko.a.n@rambler.ru
[*] Correspondence: maria.log@gmail.com

Received: 22 April 2020; Accepted: 9 May 2020; Published: 12 May 2020

Abstract: *Fagopyrum esculentum* (common buckwheat) is an important agricultural non-cereal grain plant. Despite extensive genetic studies, the information on its mitochondrial genome is still lacking. Using long reads generated by single-molecule real-time technology coupled with circular consensus sequencing (CCS) protocol, we assembled the buckwheat mitochondrial genome and detected that its prevalent form consists of 10 circular chromosomes with a total length of 404 Kb. In order to confirm the presence of a multipartite structure, we developed a new targeted assembly tool capable of processing long reads. The mitogenome contains all genes typical for plant mitochondrial genomes and long inserts of plastid origin (~6.4% of the total mitogenome length). Using this new information, we characterized the genetic diversity of mitochondrial and plastid genomes in 11 buckwheat cultivars compared with the ancestral subspecies, *F. esculentum* ssp. *ancestrale*. We found it to be surprisingly low within cultivars: Only three to six variations in the mitogenome and one to two in the plastid genome. In contrast, the divergence with *F. esculentum* ssp. *ancestrale* is much higher: 220 positions differ in the mitochondrial genome and 159 in the plastid genome. The SNPs in the plastid genome are enriched in non-synonymous substitutions, in particular in the genes involved in photosynthesis: *psbA*, *psbC*, and *psbH*. This presumably reflects the selection for the increased photosynthesis efficiency as a part of the buckwheat breeding program.

Keywords: mitochondrial genome; buckwheat; plastid genome; genetic diversity; long reads; targeted assembly

---

1. Introduction

In contrast to plastid genomes, plant mitochondrial genomes are usually large and complex. They are shaped by multiple structural changes, including rearrangements, duplications, and horizontal gene transfer (HGT) from nuclear and plastid genomes and, in rare cases, from other plant species [1,2]. This complexity leads to the under-representation of the plant mitochondrial genome sequences compared to the plastome sequences. Even in the species for which complete nuclear genomes are characterized, the information on the mitochondrial genome is often lacking or incomplete. The subject of our study, *Fagopyrum esculentum*, is an important agricultural non-cereal grain plant. Three nuclear genome assemblies of *Fagopyrum* species are available by now ([3,4], and an unpublished study, accession number GCA_004303065), as well the number of plastid genomes [5,6]. However, not a single mitogenome sequence is available. At a larger scale, *Fagopyrum* belongs to the family Polygonaceae, an isolated group within the order Caryophyllales. By now the only species of Polygonaceae with a sequenced mitochondrial genome is *Fallopia multiflora* [7]; this is not sufficient for the representation of this large and diverse group of plants. The availability of the mitochondrial genome of buckwheat is also important for practical applications. The cytoplasmic male sterility (CMS) is associated with rearrangements in the mitochondrial genome. They create chimeric open reading frames (ORF) whose products interfere with the functioning of the mitochondrial electron transfer chain and are thus toxic for cells (for a review, see [8]). The characterization of mitogenomes of CMS lines allowed the finding of events presumably responsible for the sterility in many agricultural plants (e.g., [9,10]). CMS was reported in buckwheat [11]; however, its genetic basis is unknown. Additionally, extranuclear genomes, due to their uniparental inheritance, are an important source of information on the maternal lineage of a species or cultivar. Interspecific hybridization is a promising tool for breeding as it would allow the transfer of beneficial traits (self-compatibility, resistance to abiotic stresses) from related species (*F. homotropicum*, *F. tataricum*) [12]. The characterization of extranuclear genomes of all species involved in breeding will help to trace the genealogy of the hybrids. In this study, we generated the reference genome sequence for the buckwheat mitochondrial genome and studied the diversity of the mitochondrial and plastid genome in 11 buckwheat cultivars (nine Russian, one Japanese, and one Canadian) and the ancestral subspecies, *F. esculentum* ssp. *ancestrale*.

2. Results and Discussion

2.1. Genome Assembly: Standard and Custom Tools

The initial assembly was performed using Unicycler [13], an assembly tool optimized for the circular genomes. According to Unicycler, there were 10 circular contigs with a similar read coverage that ranged from 733.16 to 1835.25 and a total length of 404 Kb. The BLAST analysis indicated that these 10 contigs are mitochondrial, based on the presence of typical plant mitochondrial genes and on high similarity with the *Fallopia multiflora* mitogenome. These results are unexpected. Despite plenty of evidence that plant mitogenomes can exist in the form of multiple circles and even non-circular forms due to intramolecular recombination mediated by the repeats [14], the single circular molecule, which includes all the subcircles (so-called master circle), is usually recovered in the genome assemblies. This is true especially for the cases where the data are not limited by the shotgun short-read libraries but include mate pair libraries and/or long read data that allow resolving of long repeats. There are several reports of bipartite mitogenomes [15,16] including in Polygonaceae [7]. The higher number of circles is much rarer, in particular a number of circles more than 10 was found only in 4 out of 307 assembled plant mitochondrial genomes deposited in NCBI GenBank (as of 21 February of 2020). All of them represent very special cases: Extremely enlarged mitogenomes in the genus *Silene* [17] and parasitic plants *Lophophytum mirabile* [1] and *Cynomorium* species [18], in which a large part of the mitogenome is acquired by HGT from their hosts. Therefore, initially, we supposed that this result could be a misassembly. To check this, we developed a new assembly tool; it is called Elloreas (ELongating LOng REad ASsembler). It is based on principles similar to NOVOPlasty [19], a seed-and-extend algorithm optimized for the assembly of plastid and mitochondrial genomes out of whole genome sequencing data. While NOVOPlasty was created for short-read assembly, Elloreas performs best with long reads, though it can work with short reads too. An important feature of Elloreas is that it indicates the presence of alternative paths of the extension (in case if they exist).

Basically, Elloreas works in the following way:

1. A user provides a starting sequence, which may be a contig from another assembly or just a random sequencing read.
2. Elloreas maps reads to the 3′ end of this sequence.
3. It finds reads that overlap the ends, such that a part of a read is mapped to the 3′ end while the other part overhangs the contig.
4. Elloreas calculates sequence consensus for these "overhanging parts".
5. It extends the contig using this consensus.

6.   It repeats all steps from "2." to "6." for this extended contig.

The work of Elloreas is regulated by multiple parameters, which can optionally be changed by the user, for example, the minimum percent identity required to map a read on a contig and the minimum number of reads supporting an extension required to extend the contig. We used contigs (identified as mitochondrial based on the BLAST search) assembled by Canu and Falcon, two widely used long-read assemblers, as starting sequences. The extension of these contigs by Elloreas showed that they correspond to circular chromosomes: After several iterations of extension, the parts on the 5′ and 3′ ends were found to be identical. Additionally, Elloreas indicated that during the extension there were no "forks", i.e., several alternative extensions supported by similar amounts of reads. This confirmed the existence of distinct circular chromosomes inferred by Unicycler. The sequences of Elloreas and Unicycler contigs were identical with one exception, an 857-bp deletion in Unicycler contig mito2. The mapping of the raw reads on Unicycler contigs showed that the variant assembled by Elloreas is the correct one. We expect that Elloreas will be useful for the assembly and assembly tests of other small genomes, whether circular or not.

2.2. Genome Structure and Gene Content

The result of the assembly was the set of 10 contigs with total length 404,063 bp. Their coverage ranges from 713 to 988× (Table 1). The coverage along all 10 contigs is rather uniform (Figure S1).

Table 1. Length and coverage of buckwheat mitochondrial chromosomes.

Number of Chromosomes	Length	Coverage	Accession Number (NCBI)
1	87,722	861	MT318702
2	71,837	839	MT318703
3	52,654	872	MT318705
4	35,904	763	MT318701
5	31,499	910	MT318704
6	28,895	864	MT318706
7	27,675	713	MT318708
8	25,181	769	MT318709
9	23,738	988	MT318710
10	18,958	727	MT318707

Moving 10 kbp from the end of each contig to its start and mapping reads to such forms of contigs also indicates a uniform coverage, implying that these contigs correspond to circular sequences. Long reads generated by the Pacific Bioscience and Oxford Nanopore Technologies platforms are a great tool for the detection of alternative structural variants in mitochondrial genomes (see, e.g., [20]). We did not identify the alternative variants at high frequency; only two variants were supported by >10% reads. The most frequent is an inversion within the chromosome mito9; it is supported by approximately 16% of reads. All other variants were found at a much lower level (see Table S1). The predominant type of the structural variation is the chromosome merge (25 out of 42 variants with frequency higher than 1%). Buckwheat mitochondrial contigs have a number of repeats; in particular the largest contig, mito1, carries a large direct repeat (~10 Kb). There is also a number of smaller repeats, both direct and inverted (see Figure 1). The repeats are known to be a hotspot for the recombination (see, for example, [21,22]). Indeed, we found that many structural variants, predominantly chromosome merges, are associated with the repeats (Table S1). This shows that the buckwheat mitochondrial genome undergoes recombination, which generates a diversity of subgenomic forms.

However, these alternative forms have a lower frequency (there are no structural variants supported by more than 50% of reads). This suggests that 10 independent circular chromosomes are the predominant form of the mitochondrial genome of *F. esculentum*. The chromosomes mito1–10 carry a complete set of genes typical for plant mitogenomes: Nine *nad* genes, two *sdh* genes, *cob*, *cox1–3*, five *atp* genes, four cytochrome c maturation factors, *matR*, *mttB*, and ribosomal protein (RP) genes (Figure 1).

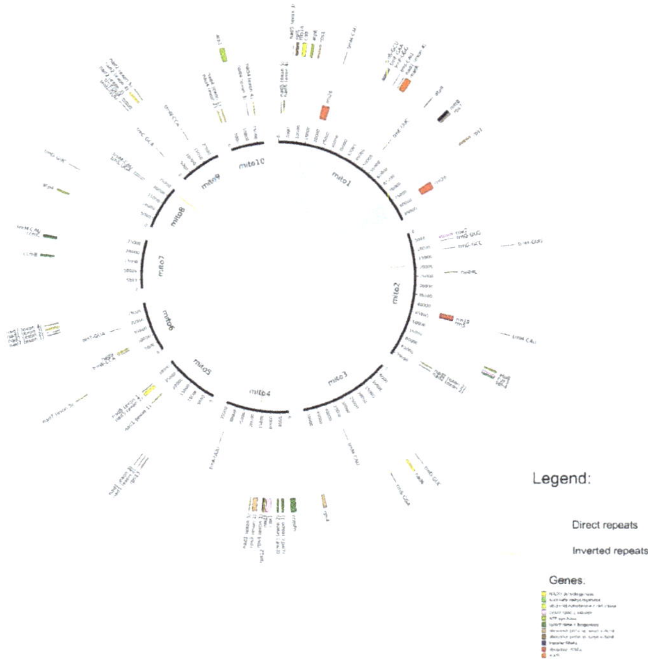

Figure 1. Map of the buckwheat mitochondrial genome representing genes and repeats. Genes shown in the outer circle are transcribed clockwise; in the inner circle, counterclockwise. Repeats with length >500 bp and similarity >95% are represented.

The latter are variable within angiosperms; virtually any of the RP genes are lost in one or more angiosperm lineages [23]. In particular, the loss of *rps13* occurred in the common ancestor of rosids and *rps8*, in the common ancestor of seed plants. Congruent with this, in *F. esculentum*, the typical mitochondrial *rps8* is absent while *rps13* is present, as well as *rps1, 3, 4, 7, 10, 12,* and *14*. Concerning large subunit RP genes, *rpl5* and *rpl16* are present while *rpl2* and *rpl10* are absent. *rpl2* is present in *Nepenthes* [24] but absent in *Fallopia* [7] and *Beta* [25], indicating multiple losses in Caryophyllales. *rpl10* is a pseudogene in *Nepenthes* and completely absent in *Beta* and *Fallopia*. It is not uncommon for mitochondrial RP genes to be replaced by their plastid counterparts, either nucleus-encoded [26] or integrated in the mitochondrial genome [27]. The *F. esculentum* mitogenome carries large insertions of DNA of plastid origin (MIPT) (see Table S2). They are especially abundant in chromosome 7 where they constitute as much as 42% of its the total length. We assume that the MIPT are recent, based on their high level of similarity with the plastome (94–100%) and on the fact that they are not shared with other Caryophyllales. The MIPTs include a number of plastid genes with intact ORFs (e.g., *rpl16, infA*) while many others have internal stop codons (*rps3, rpl22, rpl14, rps8*), making them unlikely to have functional significance. Not only the gene content but the intron content is also variable in mitochondrial genes. There are several mechanisms responsible for this, in particular, retroprocessing—the integration in the genome of the cDNA corresponding to the processed transcripts [28] and horizontal DNA transfer [29]. In *F. esculentum*, *cox1* intron, which was frequently acquired via HGT in many groups of angiosperms [30], is absent, as well as in other Caryophyllales. Five out of nine *nad* genes have introns, and three (*nad1, nad5, nad2*) of them have interchromosomal trans-splicing. In *nad1, atp6,* and *nad4L*, we found an atypical start codon ACG, suggesting RNA editing. The ACG start codon in these genes is also observed in most angiosperms, including Caryophyllales; editing converting it to a typical start codon was shown experimentally in several species [31,32]. Regarding RNA-coding genes, we found 3 rRNA genes and 24 tRNA genes. The latter divide into two groups, typical mitochondrial and chloroplast-like genes. There are 15 native mitochondrial tRNA genes (12 are unique) and 9 chloroplast-like genes (8 are unique). The genes coding for trna-Met (elongator) are duplicated.

2.3. Phylogenetic Analysis

Mitochondrial genes, due to their low substitution rate, are a valuable source of information for phylogenetic analysis at a high taxonomic level (see, e.g., [33]). Phylogenetic analysis based on a set of core mitochondrial genes in 28 species representing all major clades of angiosperms and an outgroup resulted in a tree with a topology congruent with ones based on nuclear and plastid data. In particular, Caryophyllales are resolved as a sister group to asterids and *Fagopyrum* is a sister to *Fallopia* (Figure 2).

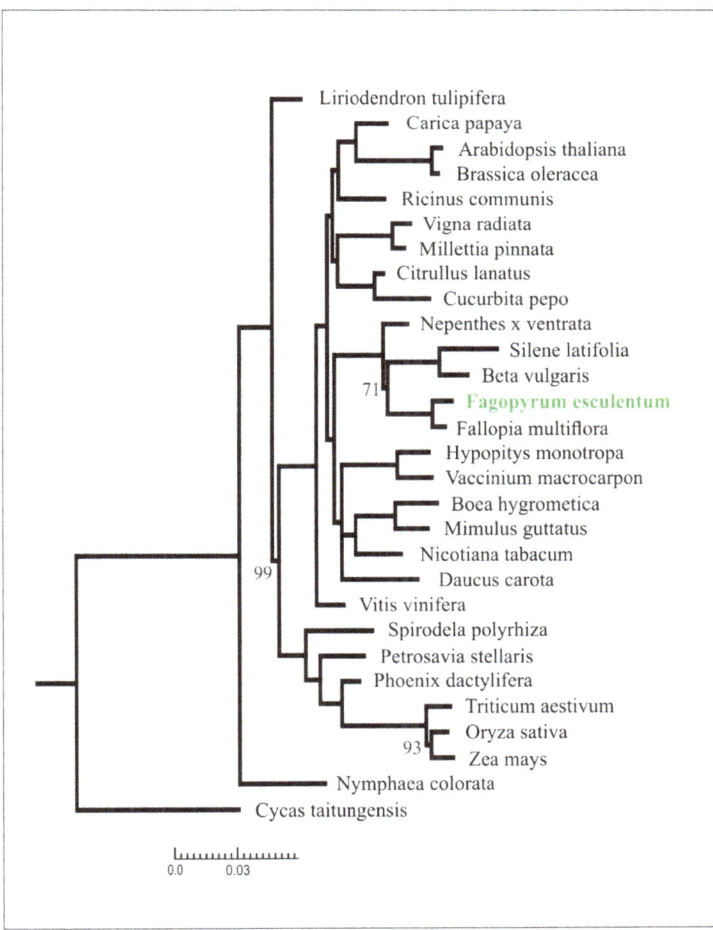

Figure 2. Phylogenetic tree based on maximum likelihood analysis of a concatenated data set of mitochondrial gene sequences. Branch lengths are proportional to the number of substitutions; bootstrap support values are equal to 100, unless otherwise specified.

While it is natural to expect the identical (or highly similar) topology of phylogenetic trees based on plastid and mitochondrial genes, this cannot be taken for granted (see, e.g., [34]). Mitochondrial genes are often acquired by horizontal gene transfer from other plant species. The most notable example is *Amborella trichopoda* [2] and parasitic plants, but the transfers of single genes and gene parts is known in many other species [35,36]. The phylogenetic analysis of single genes helps to reveal the cases of HGT via the incongruence of gene trees and species trees. In order to check for the presence of HGT, we performed phylogenetic analysis of single genes. In all of them, *Fagopyrum* was either grouped with *Fallopia* or unresolved (Figure S3); both cases are consistent with vertical transmission, thus we conclude that no buckwheat genes were subjects of HGT.

2.4. Genetic Diversity of the Mitochondrial and Plastid Genome in Common Buckwheat

The assembled sequence of the buckwheat mitogenome allowed us to estimate the diversity among cultivars. We found that it is unexpectedly low (Table S2). Only three-four substitutions differ Dasha from most other Russian cultivars. The most divergent are the cultivars Koto and Shinanonatsusoba, which have six substitutions. All variants are located in the non-coding regions. Interestingly, three out of six substitutions in Koto are shared with Russian cultivars and two more are shared with Shinanonatsusoba, the Japanese cultivar. This may indicate that Koto and Shinanonatsusoba have a recent common maternal ancestor. Shinanonatsusoba is a Japanese cultivar that originated in 1984 (Nishimaki Et al., 1984). Koto is a cultivar produced by the Canadian breeding company Kade Research in 2002. According to Ikeda [37], Canadian buckwheat breeders (not excepting Kade Research) specialized in the breeding of buckwheat for Japanese market, even the names of many Canadian cultivars (including Koto) are derived from Japanese. Thus, it is highly likely that Japanese cultivars, in particular Shinanonatsusoba, might contribute to the Koto genome. In contrast to the low divergence between cultivars, *F. esculentum* ssp. *ancestrale* differs from Dasha in 220 positions. The prevalent type of the variant is single nucleotide variation (SNV), followed by deletions and insertions (Table S2). All variants are located in the non-coding regions or in the synonymous positions of the codon and do not affect the amino acid sequence.

In an earlier study [5], we characterized the plastid genome of *F. esculentum* ssp. *ancestrale*. However, we found that it contains a number of inaccuracies, mostly indels associated with the homopolymer and low-complexity regions, which are error-prone for Sanger sequencing. Since indels in the reference genome may result in mapping and variant calling errors, we decided not to use this sequence as a reference. Taking advantage of the long reads, we assembled and annotated the plastid genome of the Dasha cultivar and used it as a backbone for the analysis of the plastid genome diversity. The diversity among cultivars is also extremely low; as well as for the mitochondrial genome, this concerns not only Russian cultivars but also Koto and Shinanonatsusoba. For Dasha and Dizajn, we found no differences at all. Dasha and other cultivars differ in one position: T to G substitution in the intron of the trnG-UCC gene. Shinanonatsusoba carries an additional change, the substitution in the trnG-trnM spacer. As well as for the mitochondrial genome, the number of SNPs with *F. esculentum* ssp. *ancestrale* is much higher: 159 (Table S4) (148 if SNPs located in the IR counted once). The SNP density is the highest in the small single copy region and the lowest in the inverted repeat (IR), as expected. In total, 81 SNPs are located in the spacers, 20 in the introns, and 58 in the coding regions. What is surprising is that 32 substitutions are non-synonymous (Table 2). While several types of mutations (C-to-T substitutions) can potentially be silenced by RNA editing (see, for example, [38]), only three substitutions out of 32 are C-to-T. Most substitutions affect highly variable genes, such as *matK*, *ycf1*, and *rpoC2*; however, there are also ones that change the amino acid sequence of highly conserved genes encoding photosystem components (*psbA*, *psbH*, *psbC*). While missenses in photosynthesis-related genes usually adversely affect photosynthesis, there are several examples of a positive effect compared to the wild type [39]. We hypothesize that this mutation(s) increases the photosynthesis efficiency and their fixation in buckwheat cultivars is the result of artificial selection. The increase of the photosynthesis efficiency has indeed been an important trend in buckwheat breeding since the middle of the 20th century. The survey of photosynthesis efficiency in different cultivars and landraces showed that in modern cultivars (developed in 1990–2010, including three sampled in our study: Demetra, Dizajn, and Devyatka), it is higher than in old cultivars and local landraces up to approximately 20% [40]. At the same time, modern cultivars have a low adaptive potential, and they are less resistant to unfavorable environmental conditions (in particular, drought) [41].

Table 2. Non-synonymous substitutions that differentiate Dasha and *F. esculentum* ssp. *ancestrale* plastid genes (substitutions located in the IR are counted once).

Reference Position	Type	Reference (Dasha)	Alternative (*F. esculentum* ssp. *ancestrale*)	Coverage	Frequency	Amino Acid Change
1354	SNV	G	T	164	100	psbA:p.Leu49Ile
2105	SNV	A	C	144	100	matK:p.Phe465Val
2809	SNV	C	T	163	99,4	matK:p.Arg230Gln
3467	SNV	C	T	155	98,7	matK:p.Asp11Asn
11370	SNV	C	A	143	100	atpF:p.Lys157Asn
19716	SNV	A	T	184	100	rpoC2:p.Phe85Ile
34532	SNV	C	A	178	100	psbC:p.Leu209Ile
46072	SNV	G	T	161	100	rps4:p.Ser148Tyr
58194	SNV	T	G	188	98,9	accD:p.Phe61Leu
58536	SNV	A	C	149	98,7	accD:p.Glu175Asp
69342	SNV	G	T	129	100	rpl20:p.Ser114Tyr
75628	SNV	T	G	203	100	psbH:p.Ser48Ala
81203	SNV	C	T	152	100	rps8:p.Ser78Asn
83622	SNV	A	G	173	100	rps3:p.Met199Thr
90367	SNV	C	A	146	93,8	ycf2:p.Gln974Lys
92722	SNV	A	G	215	83,7	ycf2:p.Ser1759Gly
110512	SNV	T	G	180	99,4	ycf1:p.Phe278Leu
111003	SNV	C	A	139	100	ycf1:p.Ser442Tyr
112085	SNV	T	C	138	100	ycf1:p.Phe803Leu
113218	SNV	T	A	131	97,7	ycf1:p.Phe1180Leu
115685	SNV	G	A	144	93,1	ndhF:p.Leu687Phe
115832	SNV	A	G	118	96,6	ndhF:p.Phe638Leu
116149	SNV	G	A	142	96,5	ndhF:p.Ala532Val
116840	SNV	C	G	214	98,6	ndhF:p.Val302Leu
118832	SNV	C	G	79	100	rpl32:p.Arg49Gly
127083	SNV	G	T	225	100	ndhA:p.Ser92Arg

Taken together, the analysis of the extranuclear genome diversity shows that buckwheat cultivars sampled in this study share a very recent common maternal ancestor. This would not be surprising for Russian cultivars, but our study also included Japanese and Canadian (presumably of Japanese origin) cultivars. This might indicate the loss of genetic diversity as a side effect of the intensification of buckwheat breeding in the last several decades. A similar situation is observed for sunflower: While a high diversity exists in wild populations, most cultivated varieties arise from a limited pool of germplasm [42,43]. This calls for better characterization of the buckwheat germplasm in order to understand the patterns and limits of its diversity (including that of extranuclear genomes). We expect that the availability of reference sequences for plastid and mitochondrial genomes will facilitate this research program.

3. Materials and Methods

3.1. Data Source, DNA Extraction, and Sequencing

For 10 buckwheat cultivars (Dasha, Dizajn, Demetra, Devyatka, Dialog, Bashkirskaya krasnostebelnaya, Karadag, Kujbyshevskaya, Kazanka Russian cultivars and Shinanonatsusoba Japanese cultivar) and *F. esculentum* ssp. *ancestrale*, the data were generated in this study. DNA was extracted using the cetyl trimethylammonium bromide (CTAB)-based method [44] from a single plant (Dasha, *F. esculentum* ssp. *ancestrale*) or a pool of five plants (other cultivars). For Dasha cultivar, sequencing was performed using a Sequel II instrument (Pacific Bioscience, Menlo Park, California, USA) at DNALink company. For other cultivars (except for Koto) and *F. esculentum* ssp. *ancestrale*, libraries were prepared using a TruSeq DNA library prep kit (Illumina, San Diego, California, USA) and sequenced on a HiSeq2000 instrument. The reads are deposited in the NCBI database under bioproject number PRJNA627307. Koto sequence data were taken from the study [4], DRA accession number DRR046985.

3.2. Read Preprocessing for De Novo Assembly

Before assembly, we performed downsampling (the reduction of the number of reads). This was done in order to reduce the amount of CPU and RAM resources required for the assembly. For PacBio data used for the de novo assembly, we found based on the k-mer frequency that the whole set of reads results in a coverage of approximately 800× for the mitochondrial genome. Then, we performed two stages of the downsampling. First, using SeqTk v.1.3. (https://github.com/lh3/seqtk), we randomly picked (command "seqtk 0.1") 253,638 reads. Second, with the command "kmer_filter" from the modified version (see below) of Stacks 2.5 [45], we removed all reads with a median copy number of k-mers less than 5 (this removes the majority of reads corresponding to single-copy nuclear regions). This operation reduced the number of reads approximately twofold—from 253,638 to 133,619. By default, Stacks removes reads where 80% of k-mers have a copy number below a given threshold. We slightly changed its source code, so the required percent of low-copy k-mers to discard a read was 50%, which means that our version of Stacks uses a median copy number of k-mers in a read as a criterion.

For Illumina data, we randomly picked 50–200 millions of paired reads. As we found empirically, by sampling and mapping different numbers of reads (5–300 millions), this number of reads is sufficient to provide >1000× coverage for the plastid and >100× for the mitochondrial genome.

3.3. Mitochondrial Genome Assembly and Assembly Check

The initial assembly was performed by Unicycler 0.4.8 [13] with the default parameters. After assembly, we estimated the mean read coverages of all contigs by mapping all reads to contigs with minimap 2.17 [46] with the option "asm20", recommended by the author of minimap for mapping PacBio CCS reads. The list of coverages for all contigs was created based on the mapping, by the command "DepthOfCoverage" from GATK 3.8 [47]. Potential mitochondrial contigs were identified based on a similarity search using BLASTN 2.9.0 [48] (e-value threshold of 10^{-3}, word size 7 bp). The following genomes were used as query: The mitochondrial genomes of *Fallopia multiflora*, the only publicly available assembled mitochondrial genome from the family Polygonaceae, *Nepenthes* × *ventrata* (*N. ventricosa* × *Nepenthes alata*) and *Arabidopsis thaliana* and the plastid genome of *F. esculentum*. The latter is necessary in order to distinguish between plastid and mitochondrial contigs and to find regions of plastid origin in the mitochondrial genome. The check of the assembly was performed using dedicated genome assembler, which we named Elloreas (abbreviation for ELongating LOng REad ASsembler). It is deposited at https://github.com/shelkmike/Elloreas. Elloreas was run with parameters "sequencing_technology hifi_pacbio", "minimum_length_of_mapped_read_part 8000", "minimum_read_similarity 99%", "contig_edge_size_to_use_for_mapping 15000". As starters for Elloreas, we used several random mitochondrial (based on the similarity search described above) contigs from assemblies created by Canu 2.0 [49] and Falcon 1.1.0 [50]. The assembly of Canu was performed with all reads, the parameter "genomesize" was set to 1.5 Gbp, which is the approximate nuclear genome size of *F. esculentum* [51]. The assembly by Falcon was done using reads with a median k-mer coverage of at least 50. The estimated genome size for Falcon was initially set to 1.5 Gbp, but the sum of produced contigs lengths was only 400 Mbp, due to the removal of many nuclear reads by Stacks. Therefore, we reran the Falcon, setting the estimate of the genome size to 400 Mbp. Other parameters of Falcon were set to the values recommended by the authors of Falcon for CCS reads.

3.4. Detection of Structural Variants

Detection of structural variants in the mitochondrial genome was performed using the approach described by [52]. It consists of two main steps: Mapping by NGMLR 0.2.8 and then structural variant detection by Sniffles 1.0.11. NGMLR was run with the option made for typical PacBio reads ("presets pacbio"), as it has no dedicated option for PacBio CCS reads.

3.5. Plastid Genome Assembly

To assemble the Dasha plastid genome, we took the largest plastid contig (found based on similarity with *F. esculentum* ssp. *ancestrale* plastome) in the Unicycler assembly described above and extended it by Elloreas, which created a circular sequence after 50 iterations of extension. Elloreas was run with parameters "sequencing_technology hifi_pacbio", "minimum_length_of_mapped_read_part 8000", "minimum_read_similarity 100%", "contig_edge_size_to_use_for_mapping 15000". To find possible misassemblies, we mapped PacBio reads by minimap2 with the "asm20" parameter and performed variant calling by FreeBayes v.1.3.0, searching for variants with a variant quality at least 40 supported by more than 50% reads. We found one misassembly in a poly-A region; that misassembly was introduced by Unicycler. An alternative variant (5 adenines instead of 4) was supported by 98.6% reads mapping to this position and we fixed the genome accordingly.

A pairwise alignment by the online version of NCBI BLAST showed that this assembly was collinear with the plastid genome of *F. esculentum* spp. *ancestrale*. The plastome sequence is deposited in NCBI Genbank under accession number MT364821.

3.6. Annotation

For the mitochondrial genome, the annotation of protein-coding genes and rRNA genes was performed based on the BLAST search of genes encoded in the mitochondrial genomes of other Caryophyllales (*Beta vulgaris*, *Nepenthes* × *ventrata*, *Fallopia multiflora*). For the plastid genome, the annotation was transferred from the sequence of *F. esculentum* ssp. *ancestrale* plastome (NCBI accession NC_010776) using GATU [53] with manual correction. Annotation of tRNAs was performed using the tRNA-scan-SE server with sequence source = other mitochondrial for mitogenome and mixed for plastid genome and search mode = default.

The maps of mitochondrial and plastid chromosomes were drawn by a custom script, which utilized Circos 0.69 [54]. Repeats in the mitochondrial genome were found by BLASTN 2.9.0 without a requirement for the maximum e-value but with requirements for the minimum length (500 bp) and percent identity (95%). MIPTs were found by the same method with an identity cut-off of 90% and length cut-off of 100 bp. Since mitochondrial genomes are known to have tRNAs of plastid origin and genes with homologs in the plastid genome (*atp*, *rrn*), only hits that do not overlap with annotated mitochondrial protein-coding or RNA genes were reported.

3.7. Alignment and Phylogenetic Analysis

Sets of individual genes were aligned using MUSCLE [55] (version 3.8.31) with default parameters. For the analysis of the combined dataset, alignments of individual genes were concatenated using custom script. Phylogenetic analysis was performed using RaXML software (version 8.2.4) with parameters "-m GTRCAT -x 123, 456 -N 100 -p 098765".

3.8. Mapping and SNP Calling

As a reference for mitochondrial and plastid genome analysis, we used mitochondrial and plastid contigs of the cultivar Dasha generated in this study. Before mapping, the Illumina reads were downsampled to 50–200 million of paired-and reads, in order to minimize the representation of the numts and nupts. The mapping was performed using CLC Genomics Workbench v.9.5.4 with the following settings: match score 1, mismatch cost 3, linear gap cost, insertion cost 3, deletion cost 3, length fraction required to be aligned 1, similarity fraction 0,97, non-specific match handling—map randomly.

SNP calling was performed using the same program with the following options: ploidy 1, ignore positions with coverage above 10,000 (this mean actually no upper limit because in all mappings the highest coverage was lower than 10,000×), ignore broken pairs, minimum coverage 20, minimum count 15, minimum frequency (%) 75, base quality filter = yes, neighborhood radius 5, minimum central quality 20, minimum neighborhood quality 15, read direction filter = yes, direction frequency (%) 30, relative read direction filter = yes, significance (%) = 1, read position filter = yes, significance (%) = 1.

The buckwheat mitochondrial genome contains large inserts of plastid origin. Due to the 10-fold higher coverage of the plastid genome compared to the mitogenome, the mapping of plastid reads on these regions will interfere with the variant calling. In order to get rid of the false variants originated from the mapping of plastid reads, we excluded all variants found in the regions of plastid origin (as defined in Table S2).

Supplementary Materials: The following are available online at http://www.mdpi.com/2223-7747/9/5/618/s1, Figure S1: Read coverage of buckwheat mitochondrial chromosomes, Figure S2: Phylogenetic trees based on ML analysis of single mitochondrial gene sequences; Table S1: Structural variants detected in the *F. esculentum* mitogenome (the variants with frequency >1% are reported) and their association with repeats; Table S2: Segments of plastid origin in the mitogenome of *F. esculentum*; Table S3: Sequence variants that differ in the mitogenome of buckwheat cultivars and *F. esculentum* ssp. *ancestrale*; Table S4: Sequence variants that differ in the plastome of Dasha cultivar and *F. esculentum* ssp. *ancestrale*.

Author Contributions: Conceptualization, M.D.L.; methodology, M.I.S., M.D.L.; software, M.I.S.; validation, M.I.S., A.S.K.; formal analysis, M.I.S., A.S.K.; investigation, M.D.L., A.A.P.; resources, A.N.F.; data curation, M.I.S., M.D.L.; writing—original draft preparation, M.D.L.; writing—review and editing, M.I.S.; visualization, M.I.S., A.A.P.; supervision, M.D.L.; project administration, M.D.L.; funding acquisition, M.D.L. All authors have read and agreed to the published version of the manuscript.

Funding: This research was funded by Russian Science Foundation, grant number 18-76-10018.

Conflicts of Interest: The authors declare no conflict of interest. The funders had no role in the design of the study; in the collection, analyses, or interpretation of data; in the writing of the manuscript, or in the decision to publish the results.

References

1. Sanchez-Puerta, M.V.; Edera, A.; Gandini, C.L.; Williams, A.V.; Howell, K.A.; Nevill, P.G.; Small, I. Genome-scale transfer of mitochondrial DNA from legume hosts to the holoparasite Lophophytum mirabile (Balanophoraceae). *Mol. Phylogenetics Evol.* **2019**, *132*, 243–250. [CrossRef]
2. Bergthorsson, U.; Richardson, A.O.; Young, G.J.; Goertzen, L.R.; Palmer, J.D. Massive horizontal transfer of mitochondrial genes from diverse land plant donors to the basal angiosperm Amborella. *Proc. Natl. Acad. Sci. USA* **2004**, *101*, 17747–17752. [CrossRef] [PubMed]
3. Zhang, L.; Li, X.; Ma, B.; Gao, Q.; Du, H.; Han, Y.; Li, Y.; Cao, Y.; Qi, M.; Zhu, Y.; et al. The Tartary Buckwheat Genome Provides Insights into Rutin Biosynthesis and Abiotic Stress Tolerance. *Mol. Plant* **2017**, *10*, 1224–1237. [CrossRef]
4. Yasui, Y.; Hirakawa, H.; Ueno, M.; Matsui, K.; Katsube-Tanaka, T.; Yang, S.J.; Aii, J.; Sato, S.; Mori, M. Assembly of the draft genome of buckwheat and its applications in identifying agronomically useful genes. *DNA Res.* **2016**, *23*, 215–224. [CrossRef] [PubMed]
5. Logacheva, M.D.; Samigullin, T.H.; Dhingra, A.; Penin, A.A. Comparative chloroplast genomics and phylogenetics of *Fagopyrum esculentum* ssp. ancestrale–A wild ancestor of cultivated buckwheat. *BMC Plant Biol.* **2008**, *8*, 59. [CrossRef] [PubMed]
6. Wang, X.; Zhou, T.; Bai, G.; Zhao, Y. Complete chloroplast genome sequence of *Fagopyrum dibotrys*: Genome features, comparative analysis and phylogenetic relationships. *Sci. Rep.* **2018**, *8*. [CrossRef]
7. Kim, C.-K.; Kim, Y.-K. The multipartite mitochondrial genome of *Fallopia multiflora* (Caryophyllales: Polygonaceae). *Mitochondrial DNA Part B* **2018**, *3*, 155–156. [CrossRef]
8. Horn, R.; Gupta, K.J.; Colombo, N. Mitochondrion role in molecular basis of cytoplasmic male sterility. *Mitochondrion* **2014**, *19*, 198–205. [CrossRef]

9. Tuteja, R.; Saxena, R.K.; Davila, J.; Shah, T.; Chen, W.; Xiao, Y.-L.; Fan, G.; Saxena, K.B.; Alverson, A.J.; Spillane, C.; et al. Cytoplasmic Male Sterility-Associated Chimeric Open Reading Frames Identified by Mitochondrial Genome Sequencing of Four Cajanus Genotypes. *DNA Res.* **2013**, *20*, 485–495. [CrossRef]

10. Makarenko, M.S.; Usatov, A.V.; Tatarinova, T.V.; Azarin, K.V.; Logacheva, M.D.; Gavrilova, V.A.; Horn, R. Characterization of the mitochondrial genome of the MAX1 type of cytoplasmic male-sterile sunflower. *BMC Plant Biol.* **2019**, *19*. [CrossRef]

11. Zheleznov, A. Kompleksnyj Podhod k Sozdaniju Ishodnogo Materiala Dlja Selekcionno-Geneticheskih Issledovanij: Na Primere Rjada Vidov Rastenij (A Complex Approach for the Development of Material for Genetics and Breeding Studies, on the Example of Several Plant Species). Ph.D. Thesis, Institute of Cytology and Genetics, Novosibirsk, Russia, 2000.

12. Mendler-Drienyovszki, N.; Cal, A.J.; Dobránszki, J. Progress and prospects for interspecific hybridization in buckwheat and the genus *Fagopyrum*. *Biotechnol. Adv.* **2013**, *31*, 1768–1775. [CrossRef] [PubMed]

13. Wick, R.R.; Judd, L.M.; Gorrie, C.L.; Holt, K.E. Unicycler: Resolving bacterial genome assemblies from short and long sequencing reads. *PLoS Comput. Biol.* **2017**, *13*, e1005595. [CrossRef] [PubMed]

14. Kozik, A.; Rowan, B.A.; Lavelle, D.; Berke, L.; Schranz, M.E.; Michelmore, R.W.; Christensen, A.C. The alternative reality of plant mitochondrial DNA: One ring does not rule them all. *PLoS Genet.* **2019**, *15*, e1008373. [CrossRef]

15. Shearman, J.R.; Sonthirod, C.; Naktang, C.; Pootakham, W.; Yoocha, T.; Sangsrakru, D.; Jomchai, N.; Tragoonrung, S.; Tangphatsornruang, S. The two chromosomes of the mitochondrial genome of a sugarcane cultivar: Assembly and recombination analysis using long PacBio reads. *Sci. Rep.* **2016**, *6*. [CrossRef] [PubMed]

16. Shtratnikova, V.Y.; Schelkunov, M.I.; Penin, A.A.; Logacheva, M.D. Mitochondrial genome of non-photosynthetic mycoheterotrophic plant Hypopitys monotropa, its structure, gene expression and RNA editing. *Biorxiv Evol. Biol.* **2019**. [CrossRef]

17. Sloan, D.B.; Alverson, A.J.; Chuckalovcak, J.P.; Wu, M.; McCauley, D.E.; Palmer, J.D.; Taylor, D.R. Rapid Evolution of Enormous, Multichromosomal Genomes in Flowering Plant Mitochondria with Exceptionally High Mutation Rates. *PLoS Biol.* **2012**, *10*, e1001241. [CrossRef]

18. Bellot, S.; Cusimano, N.; Luo, S.; Sun, G.; Zarre, S.; Gröger, A.; Temsch, E.; Renner, S.S. Assembled Plastid and Mitochondrial Genomes, as well as Nuclear Genes, Place the Parasite Family Cynomoriaceae in the Saxifragales. *Genome Boil. Evol.* **2016**, *8*, 2214–2230. [CrossRef]

19. Dierckxsens, N.; Mardulyn, P.; Smits, G. NOVOPlasty: De novo assembly of organelle genomes from whole genome data. *Nucleic Acids Res.* **2016**, gkw955. [CrossRef]

20. Kozik, A.; Rowan, B.; Lavelle, D.; Berke, L.; Schranz, M.E.; Michelmore, R.W.; Christensen, A.C. The alternative reality of plant mitochondrial DNA. *BioRxiv* **2019**. [CrossRef]

21. Palmer, J.D.; Shields, C.R. Tripartite structure of the Brassica campestris mitochondrial genome. *Nature* **1984**, *307*, 437–440. [CrossRef]

22. Alverson, A.J.; Zhuo, S.; Rice, D.W.; Sloan, D.B.; Palmer, J.D. The Mitochondrial Genome of the Legume Vigna radiata and the Analysis of Recombination across Short Mitochondrial Repeats. *PLoS ONE* **2011**, *6*, e16404. [CrossRef] [PubMed]

23. Adams, K.L.; Qiu, Y.-L.; Stoutemyer, M.; Palmer, J.D. Punctuated evolution of mitochondrial gene content: High and variable rates of mitochondrial gene loss and transfer to the nucleus during angiosperm evolution. *Proc. Natl. Acad. Sci. USA* **2002**, *99*, 9905–9912. [CrossRef] [PubMed]

24. Gruzdev, E.V.; Mardanov, A.V.; Beletsky, A.V.; Ravin, N.V.; Skryabin, K.G. The complete mitochondrial genome of the carnivorous flowering plant *Nepenthes X Ventrata*. *Mitochondrial DNA Part B* **2018**, *3*, 1259–1260. [CrossRef]

25. Kubo, T.; Nishizawa, S.; Sugawara, A.; Itchoda, N.; Estiati, A.; Mikami, T. The complete nucleotide sequence of the mitochondrial genome of sugar beet (*Beta vulgaris* L.) reveals a novel gene for tRNACys(GCA). *Nucleic Acids Res.* **2000**, *28*, 2571–2576. [CrossRef] [PubMed]

26. Kubo, N.; Arimura, S.-I. Discovery of the rpl10 Gene in Diverse Plant Mitochondrial Genomes and Its Probable Replacement by the Nuclear Gene for Chloroplast RPL10 in Two Lineages of Angiosperms. *DNA Res.* **2010**, *17*, 1–9. [CrossRef]

27. Iorizzo, M.; Senalik, D.; Szklarczyk, M.; Grzebelus, D.; Spooner, D.; Simon, P. De novo assembly of the carrot mitochondrial genome using next generation sequencing of whole genomic DNA provides first evidence of DNA transfer into an angiosperm plastid genome. *BMC Plant Biol.* **2012**, *12*, 61. [CrossRef]
28. Cuenca, A.; Ross, T.G.; Graham, S.W.; Barrett, C.F.; Davis, J.I.; Seberg, O.; Petersen, G. Localized Retroprocessing as a Model of Intron Loss in the Plant Mitochondrial Genome. *Genome Boil. Evol.* **2016**, *8*, 2176–2189. [CrossRef]
29. Sanchez-Puerta, M.V.; Cho, Y.; Mower, J.P.; Alverson, A.J.; Palmer, J.D. Frequent, Phylogenetically Local Horizontal Transfer of the cox1 Group I Intron in Flowering Plant Mitochondria. *Mol. Boil. Evol.* **2008**, *25*, 1762–1777. [CrossRef]
30. Cho, Y.; Qiu, Y.L.; Kuhlman, P.; Palmer, J.D. Explosive invasion of plant mitochondria by a group I intron. *Proc. Natl. Acad. Sci. USA* **1998**, *95*, 14244–14249. [CrossRef]
31. Onodera, Y.; Yamamoto, M.P.; Kubo, T.; Mikami, T. Heterogeneity of the atp6 Presequences in Normal and Different Sources of Male-Sterile Cytoplasms of Sugar Beet. *J. Plant Physiol.* **1999**, *155*, 656–660. [CrossRef]
32. Tsujimura, M.; Kaneko, T.; Sakamoto, T.; Kimura, S.; Shigyo, M.; Yamagishi, H.; Terachi, T. Multichromosomal structure of the onion mitochondrial genome and a transcript analysis. *Mitochondrion* **2019**, *46*, 179–186. [CrossRef] [PubMed]
33. Hiesel, R.; von Haeseler, A.; Brennicke, A. Plant mitochondrial nucleic acid sequences as a tool for phylogenetic analysis. *Proce. Natl. Acad. Sci. USA* **1994**, *91*, 634–638. [CrossRef] [PubMed]
34. Rydin, C.; Wikström, N.; Bremer, B. Conflicting results from mitochondrial genomic data challenge current views of Rubiaceae phylogeny. *Am. J. Bot.* **2017**, *104*, 1522--1532. [CrossRef] [PubMed]
35. Bergthorsson, U.; Adams, K.L.; Thomason, B.; Palmer, J.D. Widespread horizontal transfer of mitochondrial genes in flowering plants. *Nature* **2003**, *424*, 197–201. [CrossRef] [PubMed]
36. Hao, W.; Richardson, A.O.; Zheng, Y.; Palmer, J.D. Gorgeous mosaic of mitochondrial genes created by horizontal transfer and gene conversion. *Proce. Natl. Acad. Sci. USA* **2010**, *107*, 21576–21581. [CrossRef] [PubMed]
37. Ikeda, K. Buckwheat composition, chemistry, and processing. In *Advances in Food and Nutrition Research*; Elsevier: Amsterdam, The Netherlands, 2002; pp. 395–434; ISBN 978-0-12-016444-8.
38. Fiebig, A. Rapid evolution of RNA editing sites in a small non-essential plastid gene. *Nucleic Acids Res.* **2004**, *32*, 3615–3622. [CrossRef]
39. Giardi, M.T.; Rea, G.; Lambreva, M.D.; Antonacci, A.; Pastorelli, S.; Bertalan, I.; Johanningmeier, U.; Mattoo, A.K. Mutations of Photosystem II D1 Protein That Empower Efficient Phenotypes of Chlamydomonas reinhardtii under Extreme Environment in Space. *PLoS ONE* **2013**, *8*, e64352. [CrossRef]
40. Amelin, A.V.; Chekalin, E.I.; Zaikin, V.V.; Fesenko, A.N. Photosynthesis response to the changes in light intensity and CO2 concentration in leaves of buckwheat cultivars produced at different periods [Reakcija fotosinteza list"ev sortov grechihi raznyh periodov selekcii na izmenenie intensivnosti sveta i koncentracii so2 v vozduhe]. *Vestnik BelGAU* **2017**, *4*, 133–136.
41. Amelin, A.V.; Fesenko, A.N.; Chekalin, E.I.; Zaikin, V.V. Adaptiveness of productivity and photosynthesis in buckwheat (Fagopyrum esculentum Moench) landraces and varieties produced at different periods [Adaptivnij potencial fotosinteza i produkcionnogo processa u mestnyh form i sortoobrazcov grechihi (Fagopyrum esculentum Moench) raznyh periodov selekcii]. *Sel'skokhozyaistvennaya Biol.* **2016**, *51*, 79–88. [CrossRef]
42. Rieseberg, L.H.; Seiler, G.J. Molecular Evidence and the Origin and Development of the Domesticated Sunflower (Helianthus annum, Asteraceae). *Econ. Bot.* **1990**, *44*, 79–91. [CrossRef]
43. Filippi, C.V.; Merino, G.A.; Montecchia, J.F.; Aguirre, N.C.; Rivarola, M.; Naamati, G.; Fass, M.I.; Álvarez, D.; Di Rienzo, J.; Heinz, R.A.; et al. Genetic Diversity, Population Structure and Linkage Disequilibrium Assessment among International Sunflower Breeding Collections. *Genes* **2020**, *11*, 283. [CrossRef] [PubMed]
44. Doyle, J.; Doyle, J. A rapid DNA isolation procedure for small quantities of fresh leaf tissue. *Phytochem. Bull.* **1987**, 33886.
45. Catchen, J.; Hohenlohe, P.A.; Bassham, S.; Amores, A.; Cresko, W.A. Stacks: An analysis tool set for population genomics. *Mol. Ecol.* **2013**, *22*, 3124–3140. [CrossRef] [PubMed]
46. Li, H. Minimap2: Pairwise alignment for nucleotide sequences. *Bioinformatics* **2018**, *34*, 3094–3100. [CrossRef]

47. McKenna, A.; Hanna, M.; Banks, E.; Sivachenko, A.; Cibulskis, K.; Kernytsky, A.; Garimella, K.; Altshuler, D.; Gabriel, S.; Daly, M.; et al. The Genome Analysis Toolkit: A MapReduce framework for analyzing next-generation DNA sequencing data. *Genome Res.* **2010**, *20*, 1297–1303. [CrossRef] [PubMed]

48. Camacho, C.; Coulouris, G.; Avagyan, V.; Ma, N.; Papadopoulos, J.; Bealer, K.; Madden, T.L. BLAST+: Architecture and applications. *BMC Bioinform.* **2009**, *10*, 421. [CrossRef] [PubMed]

49. Koren, S.; Walenz, B.P.; Berlin, K.; Miller, J.R.; Bergman, N.H.; Phillippy, A.M. Canu: Scalable and accurate long-read assembly via adaptive *k* -mer weighting and repeat separation. *Genome Res.* **2017**, *27*, 722–736. [CrossRef]

50. Chin, C.-S.; Peluso, P.; Sedlazeck, F.J.; Nattestad, M.; Concepcion, G.T.; Clum, A.; Dunn, C.; O'Malley, R.; Figueroa-Balderas, R.; Morales-Cruz, A.; et al. Phased diploid genome assembly with single-molecule real-time sequencing. *Nat. Methods* **2016**, *13*, 1050–1054. [CrossRef]

51. Nagano, M.; Aii, J.; Campbell, C.; Kawasaki, S.; Adachi, T. Genome size analysis of the genus *Fagopyrum*. *Fagopyrum* **2000**, *17*, 35–39.

52. Sedlazeck, F.J.; Rescheneder, P.; Smolka, M.; Fang, H.; Nattestad, M.; von Haeseler, A.; Schatz, M.C. Accurate detection of complex structural variations using single-molecule sequencing. *Nat. Methods* **2018**, *15*, 461–468. [CrossRef]

53. Tcherepanov, V.; Ehlers, A.; Upton, C. Genome Annotation Transfer Utility (GATU): Rapid annotation of viral genomes using a closely related reference genome. *BMC Genom.* **2006**, *7*. [CrossRef] [PubMed]

54. Krzywinski, M.; Schein, J.; Birol, I.; Connors, J.; Gascoyne, R.; Horsman, D.; Jones, S.J.; Marra, M.A. Circos: An information aesthetic for comparative genomics. *Genome Res.* **2009**, *19*, 1639–1645. [CrossRef] [PubMed]

55. Edgar, R.C. MUSCLE: Multiple sequence alignment with high accuracy and high throughput. *Nucleic Acids Res.* **2004**, *32*, 1792–1797. [CrossRef] [PubMed]

Article

Mitochondrial Genome of *Fagus sylvatica* L. as a Source for Taxonomic Marker Development in the Fagales

Malte Mader [1,†], Hilke Schroeder [1,†], Thomas Schott [1], Katrin Schöning-Stierand [1,2], Ana Paula Leite Montalvão [1], Heike Liesebach [1], Mirko Liesebach [1], Barbara Fussi [3] and Birgit Kersten [1,*]

[1] Thünen Institute of Forest Genetics, D-22927 Grosshansdorf, Germany; malte.mader@thuenen.de (M.M.); hilke.schroeder@thuenen.de (H.S.); thomas.schott@thuenen.de (T.S.); stierand@zbh.uni-hamburg.de (K.S.-S.); ana.montalvao@thuenen.de (A.P.L.M.); heike.liesebach@thuenen.de (H.L.); mirko.liesebach@thuenen.de (M.L.)

[2] Center for Bioinformatics, Universität Hamburg, 20146 Hamburg, Germany

[3] Bavarian Office for Forest Genetics, 83317 Teisendorf, Germany; barbara.fussi@awg.bayern.de

[*] Correspondence: birgit.kersten@thuenen.de; Tel.: +49-410-269-6105

[†] These authors contributed equally.

Received: 30 July 2020; Accepted: 24 September 2020; Published: 27 September 2020

Abstract: European beech, *Fagus sylvatica* L., is one of the most important and widespread deciduous tree species in Central Europe and is widely managed for its hard wood. The complete DNA sequence of the mitochondrial genome of *Fagus sylvatica* L. was assembled and annotated based on Illumina MiSeq reads and validated using long reads from nanopore MinION sequencing. The genome assembled into a single DNA sequence of 504,715 bp in length containing 58 genes with predicted function, including 35 protein-coding, 20 tRNA and three rRNA genes. Additionally, 23 putative protein-coding genes were predicted supported by RNA-Seq data. Aiming at the development of taxon-specific mitochondrial genetic markers, the tool SNPtax was developed and applied to select genic SNPs potentially specific for different taxa within the Fagales. Further validation of a small SNP set resulted in the development of four CAPS markers specific for *Fagus*, Fagaceae, or Fagales, respectively, when considering over 100 individuals from a total of 69 species of deciduous trees and conifers from up to 15 families included in the marker validation. The CAPS marker set is suitable to identify the genus *Fagus* in DNA samples from tree tissues or wood products, including wood composite products.

Keywords: mitochondrial genome; genome assembly; *Fagus*; Fagaceae; Fagales; molecular marker; mitochondrial marker; taxon assignment; CAPS marker; SNP

---

1. Introduction

Fagus is a genus of deciduous trees in the family Fagaceae (order Fagales), native to temperate Europe, Asia, and North America. The genus is divided into two subgenera, Engleriana and Fagus [1] and comprises 11 species (without ssp., var. and f.) [2]. As a naturally growing forest tree, European beech (*F. sylvatica* L.; subgenus Fagus) is among the most important and widespread tree species in Central Europe and is widely managed for its versatile hard wood. Because of the significance of the genus *Fagus*, genomic sequence resources are highly desired for this genus.

Mishra et al. [3] published a 542 Mb nuclear draft genome sequence of an up to 300-year-old *F. sylvatica* individual (Bhaga) from an undisturbed stand in the Kellerwald-Edersee National Park in Central Germany. The assembly comprised 6451 scaffolds that are not yet assigned to any chromosomes.

Complete chloroplast genomes are publicly available for *F. sylvatica* [4], *F. engleriana* [5], *F. crenata* [6], and *F. japonica* var. *multinervis* [7]. No complete mitochondrial genome sequence is available for any *Fagus* species to date [8]. In the entire order of the Fagales, only two assembled mitochondrial genome sequences are publicly available: one from *Quercus variabilis* (unverified sequence; GenBank MN199236) [9] and one from *Betula pendula* (GenBank LT855379.1) [10]; however, this sequence is not annotated so far.

Plant mitochondrial genomes are much larger than those of animals and highly variable in size [8,11–16]. They have low mutation rates, but have such high rearrangement rates that there is virtually no conservation of synteny [17–22]. The plasticity of plant mitochondrial genomes, leading to genome expansion, arises primarily from repeat sequences (including nontandem repeats of 50 bp and up), intron expansion, and incorporation of plastid and nuclear DNA [12,15,17–19,22–28]. The mitochondrial genome sequences of angiosperms generally have one or more pairs of large nontandem repeats (interspersed repeats) that can act as sites for inter- and intramolecular recombination, leading to multiple alternative arrangements (isoforms; including subgenomic forms) within a species [19]. Although plant mitochondrial genomes are often assembled and displayed as circular maps, plant mitochondrial DNA (mtDNA) does most likely not exist as one large circular DNA molecule but mostly as a complex and dynamic collection of linear DNA with combinations of smaller circular and branched DNA molecules [12,14,15,17,19,26,29–34].

DNA barcoding is an effective technique in molecular taxonomy. Sequences suggested to be useful in DNA barcoding include mtDNA (e.g., *cox1* for animals), chloroplast DNA (e.g., *rbcL*, *matK*, *trnL-trnF*, *ndhF*, and *atpB*), and nuclear DNA (ITS and house-keeping genes, e.g., GAPDH; [35–39] among others). Chloroplast and mitochondrial genes are preferred over nuclear genes because most of the genes lack introns, and they are generally haploid [40]. Furthermore, each cell has many chloroplasts and mitochondria, and each one can contain several copies of the respective genome [41–43]. Thus, when sample tissue is limited, the chloroplast and the mitochondrion offer relatively abundant sources of DNA.

Advances in high-throughput sequencing (short read and single molecule long read sequencing) have promoted the assembly of complete DNA sequences of chloroplast and mitochondrial genomes and allowed for extending barcoding from single loci to whole genomes [44,45]. Especially, complete chloroplast genome (plastome) sequences provide valuable data sets to resolve complex evolutionary relationships in plastome phylogenies and improve resolution at lower taxonomic levels (e.g., [46,47]). Based on whole plastome alignments, genetic markers for species identification were developed in several studies (e.g., [44,48]).

The lower nucleotide substitution rate in mitochondrial compared to chloroplast genomes of land plants [17–22] often provides not enough variation at the species level, although a few mitochondrial markers for potential species differentiation were developed (e.g., [49]). However, nucleotide variants in mitochondrial genomes may provide promising targets for the development of DNA markers that are specific for higher taxonomic levels, such as genus, family or order.

European beech is often used in particle boards [50] and not always declared as it should be due to the European Timber Regulation that came into force in 2013 [51]. Although, species of the genus *Fagus* have not yet been added to the IUCN list [52] as "endangered" species (*F. longipetiolata* and *F. hayatae* classified as "vulnerable"), the specific ecosystems they form are in danger. Due to deforestation, old-growth forests are vanishing all over the world. Especially in Eastern Europe, the protected old forests consisting of oak, spruce, and beech have been declining during the last decades mostly because of human impact, i.e., mainly poor management practices and illegal logging of old valuable trees [53]. However, these forests play a key role for sustaining biodiversity and in climate change because they store a high amount of carbon for long time periods [53].

As a contribution to the preservation of valuable forest ecosystems, molecular markers can help to identify non-declared genera or species in different wood composite products. Identification of timber genera or species from solid wood products is much easier than from composite products as it may contain wood from many different genera or species. All the more important are molecular markers that are specific for a tree genus of interest ensuring a 100% classification probability ("golden markers"), and thus allowing for a doubtless identification of the related genus in wood composite products. To further increase the identification confidence in a tree genus, additional markers specific for higher taxonomic levels are highly desired. All potential taxon-specific markers should be validated in as many other tree species as possible (including species common in wood composite products).

This study aimed at sequencing and annotating the complete mtDNA sequence of a representative *F. sylvatica* individual, which was used as a source for the development of molecular markers suitable to identify the genus *Fagus*. The developed markers (for the genus *Fagus*, the family Fagaceae, and the order Fagales) provide a useful tool set to verify the declaration "genus *Fagus*" using wood from tree tissues or wood products, including composite wood.

2. Results

2.1. Assembly and Annotation of the DNA Sequence of the Complete Mitochondrial Genome of F. sylvatica L.

The reference specimen (FASYL_29) sequenced in this study was selected from a set of genotyped beech trees from a provenance trial used also in a former study [54]. It originates from the German population Gransee/Brandenburg, which is located in the center of the natural distribution range of *F. sylvatica* (see also Section 4.1).

The complete DNA sequence of the mitochondrial genome of FASYL_29 was assembled based on Illumina MiSeq reads (2 × 300 bp; 24×) and validated using long reads from nanopore MinION sequencing (3.2×; MiSeq and MinION reads are accessible at SRA PRJNA648273). For the validation of the mtDNA sequence, nanopore reads were mapped to the assembled sequence. The mapping result presented in Figure S1 shows that the mtDNA sequence is completely covered by overlapping long reads from nanopore sequencing.

The mitochondrial genome of the *F. sylvatica* individual FASYL_29 was assembled into a single DNA sequence of a total length of 504,715 bp and an average GC content of 45.8% (GenBank MT446430; Figure 1). The assembly may best fit on a circular map (Figure 1). This display is not corresponding to the physical structure of the genome in vivo where it more likely exists in different conformations ([14,26,33] among others; see introduction).

Furthermore, the *F. sylvatica* mitochondrial genome contains 32 interspersed repeats greater than 50 bp (Table S1) including two repeats greater than 300 bp: one inverted repeat of size 918 bp and one direct repeat of size 316 bp (Figure 1). One copy of the inverted repeat is located near an ancestral gene cluster consisting of the genes *rpl16*, *rps3*, and *rps19* [18] (Figure 1). In comparison, the mtDNA sequence of *Quercus variabilis* [9] contains 17 repeats greater than 50 bp and three repeats greater than 300 bp. The largest repeat is about 17.3 kbp in size. The mitochondrial genome of *Betula pendula* [10] contains 133 repeats greater than 50 bp and two repeats greater than 300 bp. The largest repeat is 474 bp long. Various fragments of the largest repeat of *Quercus variabilis* are included with high identity in the mtDNA sequence of *F. sylvatica* (52% of the repeat with about 97% identity included; File S1) and of *Betula pendula* (38% of the repeat with about 97% identity included; File S2). A comparison of all identified *F. sylvatica* repeats with the repeats in *Quercus variabilis* or *Betula pendula*, respectively, showed that several repeats are identical und many have high similarity (summary in Table S2; BlastN results in Files S3 and S4). One of the *F. sylvatica* repeats greater than 50 bp, namely repeat_11 (81 bp in length), is 100% identical to a *Quercus variabilis* repeat of the same length (repeat_12).

Chloroplast-like DNA sequence regions with more than 90% similarity to the *F. sylvatica* chloroplast genome sequence of the same individual, FASYL_29 (NC_041437.1) [4] account for about 0.72% of the FASYL_29 mitochondrial genome and are distributed among three distinct regions in the genome (region 1: 70,870–71,106 bp with 98.3% identity; region 2: 159,645–160,312 bp with 98.8% identity; region 3: 199,831–202,538 bp with 95.2% identity). These three regions were also featured by increased coverage (compared to all other regions) when mapping all available trimmed Illumina reads from

DNA-sequencing of FASYL_29 to the related mitochondrial genome sequence (data not shown). The increased coverage is due to an unspecific mapping of chloroplast DNA-derived reads in addition to mtDNA-derived reads to these regions.

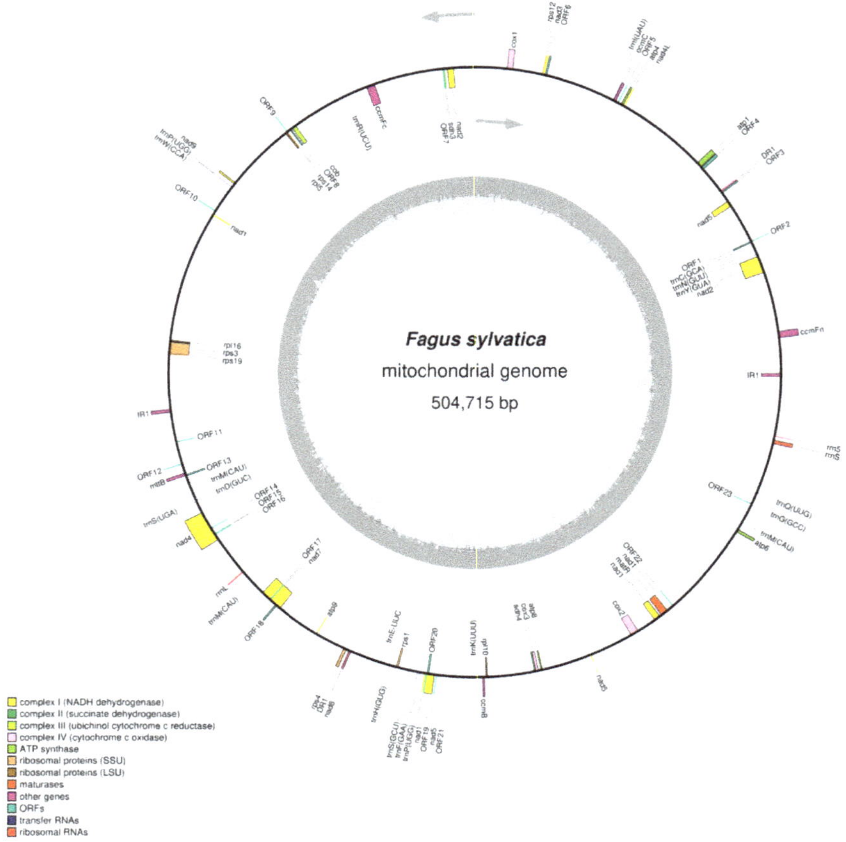

Figure 1. Circular graphical display of the assembled mitochondrial genome sequence of the *F. sylvatica* individual FASYL_29 (GenBank MT446430). The display does not correspond to the physical structure of the genome in vivo where it more likely exists in different conformations (see main text). Pairs of interspersed direct (DR) and inverted (IR) repeats longer than 300 bp and with ≥99% sequence identity are numbered (one pair each). In addition to protein-coding and structural RNA genes of predicted function, 23 potential CDS regions (indicated as "ORFs") of unknown function with support from RNA-Seq data were predicted and mapped to the genome sequence. The grey arrows indicate the direction of transcription of the two DNA strands. A GC content graph is depicted within the inner circle. The circle inside the GC content graph marks the 50% threshold. The map was created using OrganellarGenomeDraw [55,56].

In total, 58 genes with predicted function were annotated, including 35 protein-coding genes, 20 tRNA, and 3 rRNA genes. The gene *mttb* is probably a pseudogene (MT446430). All of the known genes coding for subunits of proteins of the respiratory chain were identified including *sdh3* and *sdh4* (Figure 1). Several genes coding for small or large subunits of ribosomal proteins are missing, i.e., *rps2*, *rps7*, *rps10*, *rps11*, *rps13*, *rpl2*, and *rpl15*. The genes *nad1*, *nad2*, and *nad5* were predicted to be fragmented in five exons each, belonging to more than one distinct transcription unit (MT446430).

The maturation of these genes requires cis- as well as trans-splicing events. For the following genes, more than one exon at one distinct transcription unit was predicted: *ccmFc*, *nad4*, *nad7*, *cox2*, and *rps3* (MT446430). The start codons of *nad1*, *nad4L*, and *cox1* are potentially created by RNA editing, as indicated by mappings of RNA-Seq data from two individuals of *F. sylvatica* (RNA-Seq reads accessible at SRA PRJNA648273) to the annotated mitochondrial genome sequence of *F. sylvatica* in the related regions (Figure S2). Additionally, 23 potentially protein-coding genes of unknown function were annotated based on ORF prediction from assembled RNA-Seq data (ORF1–23 in Figure 1).

In Figure S3, the gene order of potential protein-coding genes annotated in the mitochondrial genome of *Liriodendron tulipifera* [18] was compared with that of *F. sylvatica* and *Quercus variabilis* [9] as another Fagales species (*Betula pendula* was not included in the global comparison because the mitochondrial genome has not been annotated so far). As expected, there is no conservation of synteny between *Liriodendron tulipifera* and *F. sylvatica*. Although *F. sylvatica* and *Quercus variabilis* are members of the same family (Fagaceae; in different subfamilies), no larger syntenic gene groups could be identified. However, several small collinear gene clusters inferred by Richardson et al. [18] as ancestral angiosperm gene clusters in *Liriodendron tulipifera* were also identified in the *F. sylvatica* mitochondrial genome (Figure S3). Ancestral gene clusters identified in *F. sylvatica* include among others the *sdh4/cox3/atp8*-cluster (cluster also in *Quercus variabilis*), the *atp4/nad4L*-cluster (not in *Quercus variabilis*), the *cob/rps15/rpl5*-cluster including *ccmFc* (physical separation of *ccmFc* from *cob/rps15* in *Quercus variabilis*), the *rps12/nad3*-cluster (also in *Quercus variabilis*), and the *rpl16/rps3/rps19/rpl2*-cluster without *rpl2* (*rpl2* is absent in *F. sylvatica*; cluster not in *Quercus variabilis*; Figure S3).

In the comparison of the gene order (Figure S3), two gene clusters—not present in *Liriodendron tulipifera*—were identified in both *F. sylvatica* and *Quercus variabilis*: the clusters *ccmB/rpl10* and *cox1/sdh3*. The *ccmB/rpl10*-cluster was also identified in the mitochondrial genome of another Fagales member in the Betulaceae family, *Betula pendula* (Figure S4), whereas *cox1* and *sdh3* are physically separated in *Betula pendula* (draft annotation of *cox1* in File S5; *sdh3* annotation by Blast analysis). In contrast to *Betula pendula*, *F. sylvatica* and *Quercus variabilis* include the tRNA-gene *trnK* (UUU) upstream of the *ccmB/rpl10*-cluster (in a distance of about 2500 bp from *ccmB*; Figure S4). Interestingly, the *ccmB/rpl10*-cluster is not present in some non-Fagales members of the fabids analyzed, such as *Populus tremula* (NC_028096, family Malpighiales), *Vicia faba* (KC189947, Fabales; *rpl10* is not annotated), *Malus x domestica* (NC_018554, Rosales; *rpl10* is not annotated), and *Citrullus lanatus* (NC_014043, Cucurbitales; *rpl10* is not annotated).

2.2. Identification of Potentially Taxon-Specific SNPs in Mitochondrial Genes

In this study, the software SNPtax was developed allowing for the identification of potentially taxon-specific SNPs in mitochondrial genes using GenBank files of related species and outgroup species of interest as input. The software identifies SNPs specific for a pre-defined taxon based on multiple alignments of genic sequences extracted from the GenBank files together with related taxonomic information. SNPtax is freely available on https://github.com/tsciow/SNPtax.

In the search for SNPs that are potentially specific for different taxa within the Fagales, the GenBank files of the following 13 tree species were used as an input for SNPtax analysis: three Fagales species, i.e., *F. sylvatica* (GenBank Acc. MT446430; mitochondrial genome assembled and annotated in this study), *Quercus variabilis* (MN199236), and *Betula pendula* (LT855379; draft annotation in this study, see Materials and Methods) as well as 10 non-Fagales species of deciduous trees and conifers, i.e., *Bombax ceiba* (NC_038052), *Eucalyptus grandis* (NC_040010), *Lagerstroemia indica* (NC_035616), *Populus alba* (NC_041085), *Populus davidiana* (NC_035157), *Populus tremula* (NC_028096), *Populus tremula* x *Populus alba* (NC_028329), *Liriodendron tulipifera* (NC_021152), *Ginkgo biloba* (NC_027976), and *Pinus taeda* (NC_039746).

In total, 18 protein-coding genes were identified that are annotated in all 13 species. These genes were considered for the identification of potential taxon-specific SNPs. Only SNPs in conserved regions were taken into account. This way, we could identify 30 SNPs in 11 genes potentially specific for

F. sylvatica, 29 SNPs in nine genes for Fagaceae (potential Fagaceae-specific allele occurred only in the sequences of *F. sylvatica* and *Quercus variabilis*; see above), and 27 SNPs in nine genes for Fagales (specific allele only in *F. sylvatica, Quercus variabilis* and *Betula pendula*). All SNPs potentially specific for *F. sylvatica*, Fagaceae, and Fagales are summarized in Table S3. The SNPs are located in 13 different mitochondrial genes.

2.3. Development of Selected CAPS Markers and Further Validation of Their Taxon Specificity

From potentially taxon-specific SNPs (Table S3), some SNPs were selected for the development of cleaved amplified polymorphic site (CAPS) markers which fulfill the following criteria: (i) a SNP is located in a recognition site of a restriction enzyme; (ii) a SNP allele of the target taxon is part of the recognition site sequence, i.e., amplicon of the target taxon will be cut by the restriction enzyme whereas the others will not; (iii) a SNP is not located close to one of the gene ends (design of flanking primers is possible within the gene).

Specific primers were designed in the flanking regions of selected SNPs considering that the amplicon size should not exceed 200 bp because mtDNA extracted from processed wood products may be highly degraded. Moreover, primer sequences should match the respective genic region in all 13 species included in the genic alignments (see above) with a perfect match to Fagales species (as far as possible) and with a perfect match or at most one to two mismatches to the other sequences.

Selected CAPS markers were then pre-validated using DNA samples of different deciduous tree and conifer species. As an example, Figure 2 presents the pre-validation of the potential *Fagus*-specific CAPS marker 3_Fagus_*ccmFc* (marker description in Table 1). Only the five *Fagus* individuals from five different *Fagus* species included in the pre-validation provided the pattern of the digested PCR product with two fragments of 108 and 45 bp each, whereas all individuals from the other genera of deciduous trees or conifers showed the band of the non-digested PCR product of 153 bp (Figure 2).

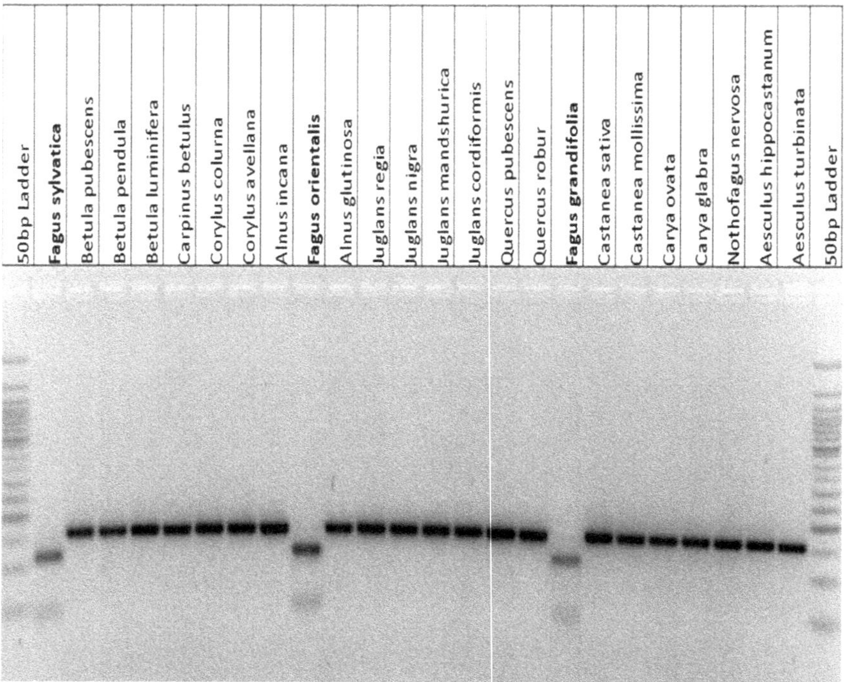

Figure 2. *Cont.*

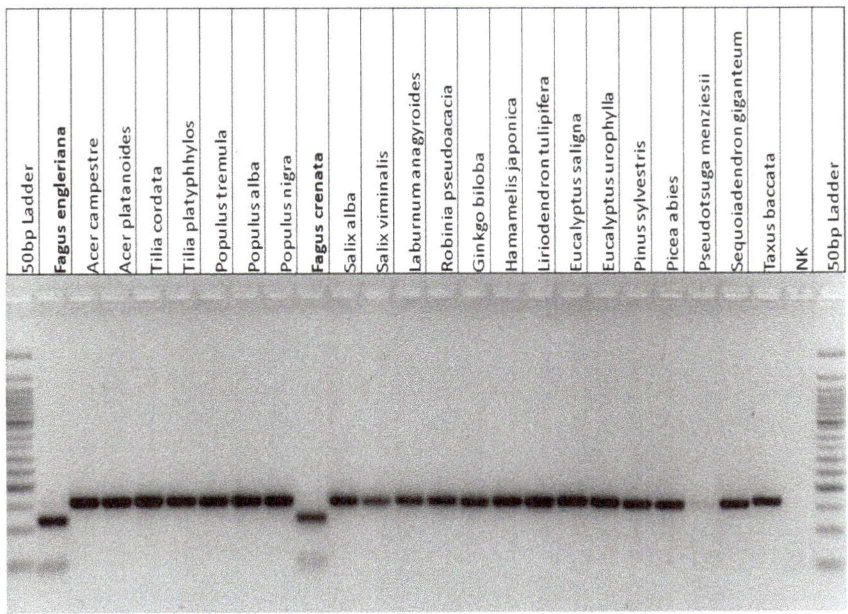

Figure 2. Pre-validation of the *Fagus*-specific mitochondrial CAPS marker 3_Fagus_*ccmFc* (marker description in Table 1) with DNA samples of *F. sylvatica*, and four other *Fagus* species as well as 41 non-*Fagus* species comprising 35 species of deciduous trees, 5 conifer species, and *Ginkgo biloba*. After digestion, PCR fragments were separated on a 1.5% agarose gel. Related primer sequences are provided in Table S4. Extended validation of CAPS markers was performed with much more individuals (see main text, Section 4.10, and Table S5).

Table 1. Mitochondrial CAPS markers specific * for genus *Fagus*, family Fagaceae, or order Fagales.

Marker Name	Taxon Specificity *	SNP Position (in bp)	Target Allele/ Alternative Alleles	PCR Amplification in Deciduous Trees (D)/ Conifers (C)	Restriction Enzyme	Fragment Sizes (Target/Others)
3_Fagus_*matR*	*Fagus*	1530	G/ C,T,A	D	*Bst*XI	53,129/ 182
3_Fagus_*ccmFc*	*Fagus*	1862	T/ C	D and C	*Bsm*BI	108,45/ 153
4_Fagaceae_*nad7*	Fagaceae	1459	A/ G	D	*Sfc*I	5786/ 143
5_Fagales_*matR*	Fagales	261	G/ A	D and C	*Nci*I	8897/ >=185

The name of the gene where the SNP is included is part of the marker name, e.g., "*matR*". SNP positions are related to the nucleotide position in the DNA sequence of the respective gene in *Populus tremula* (GenBank KT337313.1) [44] used as reference. The "target" is the taxon given in column "taxon specificity". Related primer sequences are provided in Table S4. *, The term "taxon specificity" in this context means the specificity of the related marker compared to all individuals of different deciduous tree species and conifer species included in the marker development (see above) and validation in this study (Table S5 and Section 4.10).

Only CAPS markers that successfully passed the pre-validation were subjected to extended validation as detailed below (see also Section 4.10). Table 1 summarizes the features of four successfully validated mitochondrial CAPS markers, and Figure 3 compares the related digestion patterns of these markers in one individual each of *F. sylvatica* (Fagaceae, Fagales), *Quercus robur* (Fagaceae, Fagales), *Betula pendula* (Betulacea, Fagales), and *Populus tremula* (Salicaceae, Malpighiales) as a non-Fagales species.

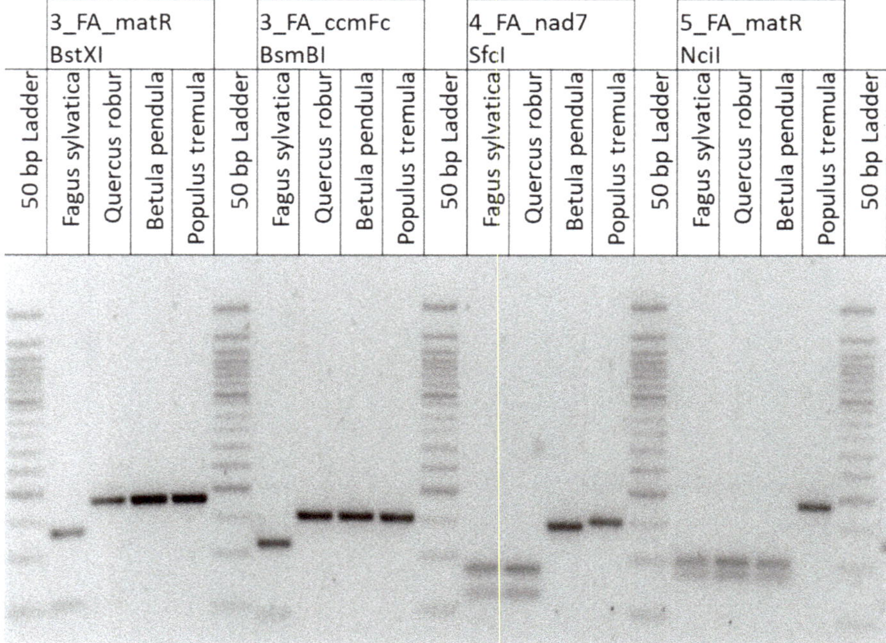

Figure 3. Representative restriction patterns of the CAPS markers (Table 1) analyzed in individuals of three Fagales and one non-Fagales species each. After digestion, PCR fragments were separated on a 1.5% agarose gel.

In the entire validation (pre-validation and extended validation), each CAPS marker (Table 1; Figure 3) was successfully validated with around 100 individuals of 59–63 tree species that belong to 14–15 families (including four Fagales families) in nine to ten orders in total (see also 4.10). All species within the entire validation, the numbers of individuals per species, and details about the origin of the individuals are summarized in Table S5. The species list contains—among others—species that are known to be potentially included in wood composite products.

Besides, the two *Fagus*-specific CAPS markers—3_Fagus_*matR* and 3_Fagus_*ccmFc* (Table 1)—were successfully validated in 25 additional *F. sylvatica*, five additional *F. orientalis*, four additional *F. grandifolia* and *F. engleriana* each, and two additional *F. crenata* individuals (Table S5). Although these two *Fagus*-specific markers were originally selected based on potentially *F. sylvatica*-specific SNPs (identified in the genic alignments described above including *F. sylvatica* as the only *Fagus* species), they turned out to be specific for the genus *Fagus* in the validation.

3. Discussion

In this study, we present the complete DNA sequence of the mitochondrial genome of *F. sylvatica* L. (Figure 1; GenBank MT446430), a deciduous tree species in the Fagaceae family. This sequence is the first complete mitochondrial genome sequence for the genus *Fagus* and the third one for the order Fagales, together with mtDNA sequences of *Quercus variabilis* (GenBank MN199236; unverified) [9] and *Betula pendula* (LT855379.1; not annotated) [10]. The mitochondrial genome sequence of the *F. sylvatica* individual FASYL_29 represents the second extranuclear genome sequence of this individual, in addition to the already published chloroplast genome sequence (NC_041437.1) [4].

Although short reads (Illumina MiSeq reads of 2 × 300 bp) were used in this study, the complete mitochondrial genome sequence could be combined from two large contigs of the initial assembly.

The assembled sequence has been successfully validated by mapping of nanopore MinION long reads (Figure S1). Because short reads encounter numerous difficulties due to low-complexity homopolymeric sequence characteristics and the potential presence of large repeat regions in mitochondrial genomes [12,17,19,22,26,42], long read sequencing is increasingly applied (often in addition to short read sequencing) for subsequent assemblies of mitochondrial genome sequences (e.g., [57–59]).

The identified size of the *F. sylvatica* mitochondrial genome assembly of 504,715 bp is between the size of the mitochondrial genomes of *Quercus variabilis*—another member of the Fagaceae family—of 412,886 bp (GenBank MN199236) [9] and *Betula pendula* (Betulaceae) of 581,505 bp (LT855379.1) [10].

Mitochondrial genomes of flowering plants are well known for their large size, fluid genome structure, and variable coding-gene set often due to horizontal gene transfer; e.g., chloroplast and nuclear sequences have been found in mitochondrial genomes or vice versa [11–13,18,19,22–28,56]. These are ongoing processes in plants. Most of the transfers in angiosperms involve ribosomal protein genes [60]. Thus, it is not unexpected that seven genes (*rps2, rps7, rps10, rps11, rps13, rpl2, rpl15*) of the ribosomal genes belonging to the ancestral gene content of the mitochondrial genome of flowering plants [18] were not annotated in *F. sylvatica* (Figure 1). Five of these seven genes (with the exception of *rps10* and *rpl2*) are also missing in the mitochondrial genome of *Quercus variabilis* (MN199236) [9]. The missing genes *rps2* and *rps11* are also lacking in the mtDNA of *Ricinus communis* [61], *Hevea brasiliensis* [62], and *Populus tremula* [44,63] among others. The absence of the *rps13* gene from mitochondrial genomes has been shown for many members of the rosids subclass [60] including *F. sylvatica* (in this study). The ribosomal gene *rps10* missing in *F. sylvatica* is also missing e.g., in *Populus tremula* [44] and *Hevea brasiliensis* [62], but present in *Ricinus communis* [61]. A loss of *rps7* was also reported in ancestors of the Fabaceae family and of *rpl2* in some Fabaceae species [64]. Although the *rpl5* gene is lacking from many of the sequenced plant mitochondrial genomes [65], it is annotated in *F. sylvatica* (MT446430). The two respiratory genes—*sdh3* and *sdh4* (encoding subunits 3 and 4 of succinate dehydrogenase)—that have been reported to be lost from the mitochondrial genome of various angiosperms [66], were annotated in *F. sylvatica* (MT446430).

Plant mitochondrial genomes have abundant interspersed repeats [12,17,19,22,26,42] often including pairs of large repeats which cause isomerization of the genome by recombination, and numerous repeats of up to several hundred base pairs that recombine only when the genome is stressed by DNA damaging agents or mutations in DNA repair pathway genes [19]. In general, the largest repeats within a species (in angiosperms often longer than about 1 kb) have been found to recombine constitutively, leading to isomerization [19]. The longest interspersed repeat in the mtDNA of *F. sylvatica* is about 1 kb (918 bp in size; Table S1) and may be responsible for isomerization. Whereas the longest repeat in the mtDNA of *Quercus variabilis* (another Fagaceae member) is 17.3 kb in size, the longest *Betula pendula* repeat is only 474 bp (Table S1). By comparing mtDNA sequences of 72 angiosperm species, Wynn and Christensen [19] found that only a part of the species (43%) shows repeats above 10 kb.

The dynamic nature of mitochondrial genomes in the Fagales is also reflected by a gene order comparison between *F. sylvatica* and *Quercus variabilis* (Figure S3) which both are members of the Fagaceae family. Although some small collinear gene clusters inferred by Richardson et al. [18] as ancestral angiosperm gene clusters in *Liriodendron tulipifera* were also identified in the mitochondrial genomes of *F. sylvatica* and *Quercus variabilis*, no larger syntenic gene groups could be identified. Interestingly, two common gene clusters—not present in *Liriodendron tulipifera*—were identified in *F. sylvatica* and *Quercus variabilis*: the clusters *ccmB/rpl10* and *cox1/sdh3*. Whether the *ccmB/rpl10*-cluster, which was also identified in *Betula pendula* (Figure S4), is a common cluster of all Fagales remains an open question for future research.

Plant mitochondria employ distinct and complex RNA metabolic mechanisms including RNA editing, splicing of group I and group II introns, maturation of transcript ends, and RNA degradation (reviewed in [34]). RNA editing (in the form of C-U base transitions) is a post-transcriptional process which is highly prevalent in mitochondria and chloroplasts of land plants [67]. Numerous C→U conversions (and in some plants also U→C) alter the coding sequences of many transcripts of the organellar genomes, while e.g., eliminating premature stop codons or creating AUG start sites, as also shown in this study for the start sites of *nad4L, cox1*, and *nad1* (Figure S2). The start codon of *cox1* is also generated by RNA editing in other land plants, e.g., *Liriodendron tulipifera, Nelumbo nucifera, Nicotiana tabacum* [68], and *Solanum tuberosum* [69]. The start codons of *nad1* and *nad4L* are also created by RNA editing in *Allium cepa, Cucumis sativus, Glycine max, Gossypium hirsutum, Liriodendron tulipifera, Nelumbo nucifera, Oryza sativa, Phoenix dactylifera*, and *Zea mays* [68], among others. In general, non-synonymous RNA editing sites were shown to be particularly highly conserved across different plant species ([68,70] among others).

Aiming at the development of mitochondrial genetic markers suitable to identify *Fagus* species from potential mixtures of different tree species in wood composite products, we sought SNPs specific for *Fagus*, Fagaceae, and Fagales in this study. In contrast to other studies that focused on classical plant barcoding regions (e.g., [71–73]), we followed a strategy similar to super-barcoding [45], however, not considering the entire mitochondrial genome, but including all mitochondrial genes common in tree species used for marker development. Because of the highly dynamic structure of mitochondrial genomes of angiosperms, alignments of complete mitochondrial genome sequences make sense only in very closely related individuals. Recombination activities involving repeated sequences may generate subgenomic forms and extensive structural variation of angiosperm mitochondrial genomes even within the same species [11,12,14,15,17,19,26,29–34].

The development of the SNPtax tool allowed us to select SNP markers potentially specific for different pre-defined taxa based on alignments of DNA sequences of mitochondrial genes (also considering intron-containing genes but excluding trans-spliced genes). The screen for taxon-specific SNPs in conserved genic regions allows considering a broad taxonomic range during the initial SNP selection and also during marker validation because primers can be designed that amplify the region of interest in tree species of various families. The developed CAPS markers (Table 1, Figure 3) are specific for the taxa *Fagus*, Fagaceae, or Fagales, respectively, when considering the tree individuals and related species (59–63 species from about 15 families and 10 orders) included in the entire validation for each marker (see also Table S5). All CAPS markers (Table 1) are located in exonic regions of the related genes with the exception of the marker 4_Fagaceae_*nad7* that is based on a Fagaceae-specific SNP in intron 2 of the *nad7* gene. An intron of the *nad7* gene (fourth intron region) was also considered in a study aiming at the identification of medicinal plants [74].

Further validation of the CAPS markers developed in this study is necessary to prove their taxon-specificity also in extended sets of tree individuals from various species, especially if they should be applied for taxon identification against a broader species background than the potential species spectrum of wood composite products. In particular, the two potentially *Fagus*-specific CAPS markers should be further validated with other *Fagus* species besides the five *Fagus* species included in this study.

Molecular markers for taxon assignment within the Fagaceae were also developed in previous studies. For species identification among common tree species of the Alps, Brunner et al. [75] developed CAPS markers based on SNPs in the intron of the plastid gene *trnL* (UAA). One of the markers allows for differentiating *F. sylvatica* from 21 other tree species tested in this study. Because no other *Fagus* species were analyzed, it is unclear if the marker is specific only for *F. sylvatica* or also for other *Fagus* species [75]. Unfortunately, an application of this marker with highly degraded DNA from processed wood products is probably not feasible (amplicon size is too large for this purpose). In another study, microsatellite primers were developed for the endangered beech tree species, *Fagus hayatae* [76]. Recently, a set of 58 SNPs has been selected from coding regions and applied for species discrimination among European white oaks [77]. Different types of molecular markers for DNA profiling of *Quercus* spp. or *Quercus* species groups were developed in other studies, e.g., based on plastid SNPs and InDels [78], short tandem repeat loci [79], or inter-primer binding sites [80].

Recent advances in real-time nanopore sequencing could pave the way to species identification using genome scale data in the future as shown in a field-based study of closely-related *Arabidopsis* species [81].

4. Material and Methods

4.1. Plant Material

The reference *F. sylvatica* specimen for DNA sequencing (FASYL_29) was selected for sequencing from a set of genotyped beech trees from a former study [54]. It originates from the German population Gransee/Brandenburg, located in the center of the natural distribution range (53°00′ N, 13°10′ O, 70 m a.s.l.) and was sampled in a progeny test. The parental population is an approved seed stand at the age of about 200 years. The selected individual is a good genetic representative of Central Europe with highest similarities to German, Czech, and Austrian beech populations.

For RNA sequencing, buds of three year old seedlings (HE1 and HE2) grown in a nursery were collected in winter. The seedlings originate from a seed lot harvested in a natural *F. sylvatica* population that is located close to Hengstberg in the Fichtel Mountains in Germany (geo-coordinates of the population: 50°07′55″; 12°11′18″).

For validation of the developed CAPS markers, leaves or buds from 145 trees of 69 different species and various geographic origins were used. A part of this plant material was sampled in the Arboretum of the Thünen Institute of Forest Genetics in Großhansdorf. Additionally, DNA from specimens of *F. sylvatica* was provided by the projects "GenMon" and "Herkunft Europa" (subproject/project conducted at the Thünen Institute of Forest Genetics). The *Quercus* specimens were provided by the previous project "DBU Weisseiche" and by North American, Russian, and Korean institutions. Different Botanical Gardens provided us also with specimens of other species (Table S5). For the sampling strategy, we always tried to obtain material from different regions of the natural distribution areas of the tree species. Thus, e.g., especially for *F. sylvatica*, individuals from all over Europe were used (Table S5). The use of a broad range of species within the orders and families was particularly important to select "golden" markers with high specificity based on validation.

4.2. DNA Sequencing of F. sylvatica L.

Dormant buds of the *F. sylvatica* individual FASYL_29 were sampled, green tissues were dissected and DNA was extracted for short read sequencing following a slightly modified ATMAB protocol based on the protocol of Dumolin et al. [82]. Standard genomic library preparation and 300 bp paired-end sequencing was performed on Illumina MiSeq at 24× coverage (GATC Biotech AG, Konstanz, Germany).

For long read sequencing, DNA was extracted using a combination of a lysis buffer, Sera-Mag Speed beads, and a purification step, adapted from [83]. Briefly, the outer layer of leaf buds was removed; then, they were cut into pieces (to facilitate grinding), collected into 2 mL Eppendorf tubes, and frozen in liquid nitrogen (10 buds in total; two buds per tube). Samples were ground using a Retsch Mill (Retsch MM300, Haan, Germany), with two stainless steel beads (5 mm) per tube, at a speed of 25 Hz for 35 s. DNA lysis was performed by adding 700 µL lysis buffer (1% PVP 40 (*w/v*), 1% PVP 10 (*w/v*), 500 mM NaCl, 100 mM Tris-HCl pH 8.0, 50 mM EDTA, 5 mM DTT, 1.25% SDS (*w/v*), and 1% Sodium metabisulfite (*w/v*)) to each of the samples, which were then mixed gently by flicking and incubated at 64 °C for 30 min. Following that, 1 µL RNase A (10 mg/mL) per 1 mL lysis buffer was added, and the samples were incubated at 37 °C for 50 min at 400 rpm on a thermomixer. After the first 20 min, 10 µL Proteinase K (800 units/mL) was added to each sample. Once the incubation time ended, samples were left on ice for 2 min to cool down, and 0.3 volume 5 M potassium acetate pH 7.5 was added. Samples were manually mixed by inversion 20 times, slowly, then centrifuged at 8000× *g* for 12 min at 4 °C.

For DNA size selection, the supernatant was transferred to clean 1.5 mL LoBind Eppendorf tubes and 0.8 V of a homogenized Sera-Mag Speed beads solution (10 mM Tris-HCl, 1 mM EDTA pH 8.0,

1.6 M NaCl, 11% PEG 8000, 0.4% beads (*v/v*)) was added, and the tubes were mixed gently by flicking. The samples were placed on a rotor for 10 min, briefly centrifuged, and placed on a magnet. Once the beads were on the back of the tubes and the solution became clear, the supernatant was discarded, and the beads were washed twice with 1 mL freshly prepared 70% ethanol. After the last ethanol wash, the tubes were removed from the magnet and briefly centrifuged. After placing them back on the magnet, the last drops of ethanol were pipetted off. The beads were air dried for 30 s; then, the tube was removed from the magnet, and 50 µL pre-heated (50 °C) 10 mM Tris-HCl pH 8.0 was added for elution. The tubes were flicked to resuspend the beads then incubated for 10 min at room temperature. Finally, the tubes were placed back on the magnet, and once the solution was clear, it was transferred to fresh tubes.

The DNA purification step was performed using chloroform:isoamylalcohol (24:1). Since the extraction was performed using multiple tubes, the eluted DNA from each tube (~80 µL) was pipetted into a single tube, comprising a total of 400 µL. Then, one volume of chloroform:isoamylalcohol was added, and it was mixed by inversion for 5 min on a rotor. After that, the tube was centrifuged at 5000× *g* for 2 min at room temperature, and the upper phase was transferred to a fresh tube. The chloroform:isoamylalcohol step was repeated, and after that, a 0.1 volume of 100% cold ethanol was added, and the sample was centrifuged at 5000× *g* for 2 min at room temperature. The pellet was washed with 70% ethanol and resuspended in 50 µL 10 mM Tris-HCl pH 8.0 for 2 h at room temperature. The sample was stored at 4 °C until library preparation.

The sequencing library was prepared using the Ligation Sequencing Kit (SQK-LSK109) following the manufacturer's instructions (Oxford Nanopore Technologies, Oxford, UK), with 2 µg DNA as input. The R9.4.1 MinION Flow Cell was primed with the Flow Cell Priming Kit (EXP-FLP002). Sequencing was performed on a MinION Mk1B device (MIN-101B) connected to a MinIT computer (MNT-001). We used the MinIT software version 19.01.1, and further basecalling was performed using guppy v3.2.2.

4.3. RNA Sequencing of Two F. sylvatica L. Individuals

Bud samples from three-year-old *F. sylvatica* seedlings (HE1 and HE2) growing in a nursery were collected and immediately put on dry ice. RNA extraction and all further steps were performed by IGA Technology Service in Udine (Udine, Italy). Libraries were prepared using Illumina TruSeq mRNA-seq Kit. Clusters were generated on a flowcell by cBot and sequenced on a HiSeq2000 by using standard Illumina sequencing workflow.

4.4. Assembly and Scaffolding of the Illumina MiSeq Reads from DNA Sequencing

Reads were trimmed with Trimmomatic version 0.36 [84] and assembled using the CLC Genomics Workbench (CLC-GWB) Version 10.1.1 (CLC-bio, a Qiagen company; Aarhus, Denmark), (length fraction = 0.9, similarity fraction = 0.95, map reads back to contigs, word size = 45). Two out of 279,689 contigs were identified as mitochondrial sequences based on high mapping coverage and comparison against the NCBI nucleotide collection database. Blastn was used to join the mitochondrial contigs by finding similarities between the sequence endings in order to create the complete DNA sequence of the mitochondrial genome. To verify the assembled sequence, we used the tool "ROUSFinde1_1", described by Wynn and Christensen [19], with default parameters to identify repeat structures in the mitochondrial sequence of *F. sylvatica* (MT446430). Accordingly, we identified repeat structures in *Quercus variabilis* (MN199236) and *Betula pendula* (LT855379.1) for the comparison of repeats between these three mitochondrial genomes (Table S1).

4.5. Mapping of Nanopore MinION Reads to the DNA Sequence of the Mitochondrial Genome of F. sylvatica

MinION reads were error-corrected using the "Correct PacBio reads 1.1." tool of the Genome Finishing Module of CLC-GWB v12.0 (coverage percentage of reads to correct = 40). Corrected reads of a length above 10,000 bp were mapped onto the final assembly of the mitochondrial genome of *F. sylvatica* by the "Map reads to contigs 1.3" tool of CLC-GWB v20.04 using default parameters with increased length and similarity fraction (length fraction = 0.98; similarity fraction = 0.95). The created stand-alone read mapping is presented in Figure S1.

4.6. Assembly of the RNA-Seq Data

RNA-Seq data from the samples HE1 and HE2 were used for the annotation of additional ORFs. Reads were trimmed with Trimmomatic version 0.36 [84]. The combined trimmed reads of HE1 and HE2 were mapped with the STAR RNA-Seq aligner to the mitochondrial reference sequence of *F. sylvatica*. This mapping was used for a reference-guided assembly of the RNA-Seq data with the software Trinity version 2.8.5 [85] which resulted in 229 mitochondrial transcript contigs.

4.7. Annotation of the DNA Sequence of the Mitochondrial Genome of F. sylvatica

Structural and functional annotation was performed using the GeSeq server [86] with default settings and NC_028096 (mtDNA sequence of *Populus tremula*) set as reference. Using the Sequin tool v13.05 [87], these draft annotations were corrected where necessary, guided by alignments to other well-characterized eudicot mtDNA sequences including those of *Arabidopsis thaliana* (NC_037304.1) [88], *Cucurbita pepo* (NC_014050.1) [65], and *Quercus variabilis* (MN199236) [9]. Additional ORFs including the ORF of the gene *rpl10* were identified by ORF prediction in transcript contigs assembled from RNA-Seq data of two individuals (see Section 4.6) using the "Find open reading frames"—tool of CLC Genomics Workbench version 20.0.3 with the following parameters: Start codons = AUG; Stop codon included in annotation = Yes; Annotate sequences = Yes; Both strands = Yes; Open-ended sequence = No; Genetic code = 1 Standard; Minimum length (codons) = 100. In total, 90 ORFs with a minimal length of 300 bp (including start and stop codons) were identified and gave rise to 24 additional annotations (ORF 1–23 and *rpl10*), which were manually added to the Seqin file (annotated as "hypothetical proteins"). Nested ORFs were not considered. If BlastP analysis of the amino acid sequence of a related ORF versus NCBI non-redundant proteins provided hits with at least 80% query coverage and 90% identity to any known protein sequence; then, this information was included into the protein description (as a comment) in the Seqin file.

To compare and visualize the gene order between different mitochondrial genomes, the web version of the software tool geneCo (gene Comparison) [89,90] was applied (using the "map comparison" option with default parameters, but displaying names of all genes; considering only potential protein-coding genes and excluding trans-spliced genes).

4.8. Multiple Alignments of Mitochondrial Gene Sequences and Selection of SNPs Specific for Pre-Defined Taxa

SNPs specific for selected taxa were identified based on the newly developed software SNPtax which uses a set of custom scripts based on BioPerl [91] and PRANK v.170427 [92] as alignment tool. By extracting the taxonomic information together with the genic sequences from different mitochondrial genomes in GenBank format and then generating multiple alignments for each gene—at every position in the alignment—the taxon can be determined for which a certain base is characteristic. The resulting output can then be screened for selected taxa of interest to retrieve taxon-specific SNPs.

In total, the complete mitochondrial genome sequence of 13 tree species (see Section 2.2) were included in the analysis presented in this study. Because the *Betula pendula* genome (LT855379) was only available without annotation, the sequence was submitted to the GeSeq server [86,93] for gene calling and functional annotation with NC_028096 as a reference. The resulting GenBank file was used as input for the SNPtax analysis beside the GenBank files of the other 12 species.

4.9. DNA Preparation, PCR Amplification, and Agarose Gel Electrophoresis for CAPS Markers

Either buds, leaves, or cambium were prepared for extraction of total DNA following a slightly modified ATMAB protocol according to Dumolin et al. [82]. The type of tissue was dependent on what was provided as reference material used in this study.

PCR amplification was performed in 20 µL volume with 20 ng DNA. For all markers shown in Table 1, the reaction mixture contains 1× BD Buffer, 2 mM MgCl$_2$, 200 µM each dNTP, 1xDMSO (NEB), 0.2 µM of each Primer, and 1 unit Taq Polymerase (DCS Pol, DNA Cloning Service, Hamburg), except for marker 5_Fagales_*mat*R (1.5 unit Taq polymerase was used). The PCR was performed with the following program: 94 °C for 4 min, followed by 35 cycles with 94 °C for 45 s, 58 °C for 1 min, 72 °C for 1 min, and additional 5 min at 72 °C final elongation at the end.

For the restriction analyses for each marker, 10 µL PCR product was used in a volume of 20 µL. The restriction analyses (in deviation from the manufacturer's protocols) were performed as follows: the PCR product amplified with the marker 5_Fagales_*mat*R was digested with 4 units of the enzyme *Nci*I for 8 h at 37 °C and 20 min of inactivation at 80 °C—the same conditions were used for 3_Fagus_*mat*R using the enzyme *Bst*XI; 4_Fagaceae_*nad7* was digested with the enzyme *Sfc*I with the same conditions as the former; 3_Fagus_*ccmFc* was digested with 5 units of the enzyme *Bsm*BI for 8 h at 55 °C and 20 min of inactivation at 80 °C. Restriction products were visualized relative to a 50 bp ladder (Life technologies, Germany, Martinsried) using a 1.5% agarose gel stained with Roti-Safe Gelstain (Carl Roth, Germany, Karlsruhe).

4.10. Validation of the CAPS Markers

The Fagales-specific marker 5_Fagales_*mat*R was tested for "Fagales-specificity" using four families with ten genera, 41 species, and 75 individuals within the order Fagales, and additionally, nine other orders including 11 families, 17 genera, and 25 species with one individual each (Table S5). For validation of the Fagaceae-specificity of the marker 4_Fagaceae_*nad7*, three genera, and 23 species with 57 individuals from the family Fagaceae, and three further families within the order Fagales using seven genera and 15 species were used. Additionally, outside the Fagales, nine orders including 11 families, 17 genera, and 26 species and individuals were tested for validation of this family-specific marker. For validation of both *Fagus*-specific markers—3_Fagus_*mat*R and 3_Fagus_*ccmFc*—58 or 60 *Fagus* individuals were used, respectively. Outside from the genus *Fagus*, further 76/79 individuals from 55/57 species in 25 genera, 13 families, and eight orders were tested (Table S5).

Supplementary Materials: The following are available online at http://www.mdpi.com/2223-7747/9/10/1274/s1, Figure S1: validation of the mitochondrial genome sequence of *F. sylvatica* (MT446430) with long Nanopore MinION reads by mapping of corrected reads (>10,000 bp) to the final assembly (Fasyl_mt: mtDNA of *F. sylvatica*; mapped reads in green: forward reads; mapped reads in read: reverse reads). Figure S2: creation of start codons (related triplet underlined) by RNA editing in transcripts of *nad4L, cox1*, and *nad1* detected in mappings of RNA-Seq data from *F. sylvatica* to the mitochondrial genome (MT446430). Figure S3: comparison of the gene order in the mitochondrial genomes of *Liriodendron tulipifera* (NC 021152), *F. sylvatica* (MT446430), and *Quercus variabilis* (MN199236) using geneCo ([89,90]; only potential protein-coding genes without trans-splicing were compared). Figure S4: The *ccmB/rpl*10-gene cluster in the mtDNA of three Fagales species (detail enlargement of the visualization of GenBank files using CLC-GWB; *F. sylvatica*, MT446430; *Quercus variabilis*, MN199236, and draft annotation of *Betula pendula*, LT855379.1 in File S4; *: reverse complement of the annotated *Betula pendula* mtDNA sequence is shown to obtain a uniform presentation of the gene cluster). Table S1: lists of interspersed repeats in the mtDNA sequences of *F. sylvatica* (MT446430), *Quercus variabilis* (MN199236), and *Betula pendula* (LT855379.1) identified using the tool "ROUSFinde1_1" [26]. Table S2: summary of results of BlastN analyses of the DNA sequences of all *F. sylvatica* interspersed repeats versus *Quercus variabilis* or *Betula pendula* repeats, respectively (BlastN details in Files S3 and S4; detailed information to the repeats in Table S1). Table S3: mitochondrial SNPs with target alleles that are potentially specific for *F. sylvatica*, Fagaceae, or Fagales, respectively (SNPs used for CAPS markers highlighted in yellow; based on multiple alignments of DNA sequences of 18 mitochondrial genes from 13 species of deciduous trees and conifers; SNP positions related to *Populus tremula* genes; GenBank NC_028096). Table S4: primer sequences for mitochondrial CAPS markers developed in this study (see Table 1). Table S5: details on individuals used for validation of the four mitochondrial CAPS markers developed in this study (see Table 1). File S1: result of BlastN analysis of the DNA sequence of the largest repeat from *Quercus variabilis* to the mtDNA sequence of *F. sylvatica* (MT446430). File S2: result of BlastN analysis of the DNA sequence of the largest repeat from *Quercus variabilis* to the mtDNA sequence of *Betula pendula* (LT855379.1). File S3: results of BlastN analyses of the DNA sequences of all *F. sylvatica* repeats versus *Quercus variabilis* repeats. File S4: results of BlastN analyses of the DNA sequences of all *F. sylvatica* repeats versus *Betula pendula* repeats. File S5: draft—GenBank file of the *Betula pendula* mitochondrial genome created by draft annotation of the mtDNA sequence LT855379.1 using GeSeq [86,93].

Author Contributions: Conceptualization, B.K. and H.S.; methodology, M.M., H.S., T.S., K.S.-S., and B.K.; software, M.M., T.S., and K.S.-S.; validation, H.S. and B.K.; formal analysis, M.M., T.S., K.S.-S., H.S., and B.K.; investigation, M.M., H.S., B.K., T.S., K.S.-S., A.P.L.M., H.L., M.L., and B.F.; resources, H.L. and M.L.; data curation, M.M., T.S., K.S.-S., and A.P.L.M.; writing—original draft preparation, B.K, M.M., H.S., T.S., A.P.L.M., and B.F.; writing—review and editing, all authors; visualization, B.K. and H.S.; supervision, B.K., H.S., H.L., M.L., and B.F.; project administration, B.K., H.S., H.L., and B.F.; funding acquisition, B.K., H.S., B.F., and H.L. All authors have read and agreed to the published version of the manuscript.

Funding: This research was funded by the German Federal Environmental Foundation, grant number 33949/01 (project "Wood DNA barcoding"); by the FEDERAL Ministry of Food and Agriculture, and the Federal Ministry for the Environment, Nature Conservation and Nuclear Safety in the research program "Waldklimafonds", grant number 28W-C-4-092-10 (project "GenMon"); by the BAVARIAN MINISTRY FOR FOOD, AGRICULTURE AND FORESTRY, grant number P31; and by core funding of the THÜNEN INSTITUTE.

Acknowledgments: We are very grateful to Stefanie Palczewski and Marie-Fee August for technical assistance. We also thank the Morton Arboretum and the North Carolina State University for providing samples of North American oaks. For the Asian oak species, we thank the Bashkirian State Agrarian University and the Korea National Arboretum. Furthermore, we thank the Botanical Gardens of Bayreuth, Bochum, Eberswalde, Goettingen, Marburg, and our former colleague Lasse Schindler who provided us with material of different species.

Conflicts of Interest: The authors declare no conflict of interest.

References

1. Renner, S.S.; Grimm, G.W.; Kapli, P.; Denk, T. Species relationships and divergence times in beeches: New insights from the inclusion of 53 young and old fossils in a birth-death clock model. *Philos. Trans. R. Soc. Lond. B Biol. Sci.* **2016**, *371*, 20150135. [CrossRef] [PubMed]

2. The Plant List. Available online: http://www.theplantlist.org/ (accessed on 1 July 2020).

3. Mishra, B.; Gupta, D.K.; Pfenninger, M.; Hickler, T.; Langer, E.; Nam, B.; Paule, J.; Sharma, R.; Ulaszewski, B.; Warmbier, J.; et al. A reference genome of the European beech (*Fagus sylvatica* L.). *Gigascience* **2018**, *7*, giy063. [CrossRef] [PubMed]

4. Mader, M.; Liesebach, H.; Liesebach, M.; Kersten, B. The complete chloroplast genome sequence of *Fagus sylvatica* L. (Fagaceae). *Mitochondrial DNA Part B* **2019**, *4*, 1818–1819. [CrossRef]

5. Yang, Y.; Zhu, J.; Feng, L.; Zhou, T.; Bai, G.; Yang, J.; Zhao, G. Plastid Genome Comparative and Phylogenetic Analyses of the Key Genera in Fagaceae: Highlighting the Effect of Codon Composition Bias in Phylogenetic Inference. *Front. Plant Sci.* **2018**, *9*, 82. [CrossRef]

6. Worth, J.R.P.; Liu, L.; Wei, F.J.; Tomaru, N. The complete chloroplast genome of *Fagus crenata* (subgenus Fagus) and comparison with *F. engleriana* (subgenus Engleriana). *PeerJ* **2019**, *7*, e7026. [CrossRef]

7. Park, J.S.; Jin, D.P.; Park, J.W.; Choi, B.H. Complete chloroplast genome of *Fagus multinervis*, a beech species endemic to Ulleung Island in South Korea. *Mitochondrial DNA Part B* **2019**, *4*, 1698–1699. [CrossRef]

8. Organelle Genome Resources at NCBI. Available online: http://www.ncbi.nlm.nih.gov/genome/organelle/ (accessed on 1 July 2020).

9. Bi, Q.X.; Li, D.X.; Zhao, Y.; Wang, M.K.; Li, Y.C.; Liu, X.J.; Wang, L.B.; Yu, H.Y. Complete mitochondrial genome of *Quercus. variabilis*. (Fagales, Fagaceae). *Mitochondrial DNA Part B* **2019**, *4*, 3927–3928. [CrossRef]

10. Salojarvi, J.; Smolander, O.P.; Nieminen, K.; Rajaraman, S.; Safronov, O.; Safdari, P.; Lamminmaki, A.; Immanen, J.; Lan, T.Y.; Tanskanen, J.; et al. Genome sequencing and population genomic analyses provide insights into the adaptive landscape of silver birch. *Nat. Genet.* **2017**, *49*, 904–912. [CrossRef]

11. Gualberto, J.M.; Newton, K.J. Plant Mitochondrial Genomes: Dynamics and Mechanisms of Mutation. *Annu. Rev. Plant Biol.* **2017**, *68*, 225–252. [CrossRef]

12. Morley, S.A.; Nielsen, B.L. Plant mitochondrial DNA. *Front. Biosci. Landmrk* **2017**, *22*, 1023–1032.

13. Ward, B.L.; Anderson, R.S.; Bendich, A.J. The Mitochondrial Genome Is Large and Variable in a Family of Plants (Cucurbitaceae). *Cell* **1981**, *25*, 793–803. [CrossRef]

14. Bendich, A.J. Structural analysis of mitochondrial DNA molecules from fungi and plants using moving pictures and pulsed-field gel electrophoresis. *J. Mol. Biol.* **1996**, *255*, 564–588. [CrossRef] [PubMed]

15. Oldenburg, D.J.; Bendich, A.J. DNA maintenance in plastids and mitochondria of plants. *Front. Plant Sci.* **2015**, *6*, 883. [CrossRef]

16. Bendich, A.J. The size and form of chromosomes are constant in the nucleus, but highly variable in bacteria, mitochondria and chloroplasts. *BioEssays* **2007**, *29*, 474–483. [CrossRef]

17. Palmer, J.D.; Herbon, L.A. Plant Mitochondrial-DNA Evolves Rapidly in Structure, but Slowly in Sequence. *J. Mol. Evol.* **1988**, *28*, 87–97. [CrossRef] [PubMed]
18. Richardson, A.O.; Rice, D.W.; Young, G.J.; Alverson, A.J.; Palmer, J.D. The "fossilized" mitochondrial genome of *Liriodendron tulipifera*: Ancestral gene content and order, ancestral editing sites, and extraordinarily low mutation rate. *BMC Biol.* **2013**, *11*, 29. [CrossRef] [PubMed]
19. Wynn, E.L.; Christensen, A.C. Repeats of Unusual Size in Plant Mitochondrial Genomes: Identification, Incidence and Evolution. *G3-Genes Genom. Genet.* **2019**, *9*, 549–559. [CrossRef]
20. Drouin, G.; Daoud, H.; Xia, J. Relative rates of synonymous substitutions in the mitochondrial, chloroplast and nuclear genomes of seed plants. *Mol. Phylogenet. Evol.* **2008**, *49*, 827–831. [CrossRef]
21. Wolfe, K.H.; Li, W.H.; Sharp, P.M. Rates of Nucleotide Substitution Vary Greatly among Plant Mitochondrial, Chloroplast, and Nuclear Dnas. *Proc. Natl. Acad. Sci. USA* **1987**, *84*, 9054–9058. [CrossRef]
22. Schuster, W.; Brennicke, A. The Plant Mitochondrial Genome-Physical Structure, Information-Content, Rna Editing, and Gene Migration to the Nucleus. *Annu. Rev. Plant. Phys.* **1994**, *45*, 61–78. [CrossRef]
23. Bullerwell, C.E.; Gray, M.W. Evolution of the mitochondrial genome: Protist connections to animals, fungi and plants. *Curr. Opin. Microbiol.* **2004**, *7*, 528–534. [CrossRef] [PubMed]
24. Bock, R. The give-and-take of DNA: Horizontal gene transfer in plants. *Trends. Plant Sci.* **2010**, *15*, 11–22. [CrossRef] [PubMed]
25. Kleine, T.; Maier, U.G.; Leister, D. DNA transfer from organelles to the nucleus: The idiosyncratic genetics of endosymbiosis. *Annu. Rev. Plant Biol.* **2009**, *60*, 115–138. [CrossRef] [PubMed]
26. Kozik, A.; Rowan, B.A.; Lavelle, D.; Berke, L.; Schranz, M.E.; Michelmore, R.W.; Christensen, A.C. The alternative reality of plant mitochondrial DNA: One ring does not rule them all. *PLoS Genet.* **2019**, *15*, e1008373. [CrossRef] [PubMed]
27. Dong, S.; Chen, L.; Liu, Y.; Wang, Y.; Zhang, S.; Yang, L.; Lang, X.; Zhang, S. The draft mitochondrial genome of *Magnolia biondii* and mitochondrial phylogenomics of angiosperms. *PLoS ONE* **2020**, *15*, e0231020. [CrossRef] [PubMed]
28. Adams, K.L.; Palmer, J.D. Evolution of mitochondrial gene content: Gene loss and transfer to the nucleus. *Mol. Phylogenet. Evol.* **2003**, *29*, 380–395. [CrossRef]
29. Tsujimura, M.; Kaneko, T.; Sakamoto, T.; Kimura, S.; Shigyo, M.; Yamagishi, H.; Terachi, T. Multichromosomal structure of the onion mitochondrial genome and a transcript analysis. *Mitochondrion* **2019**, *46*, 179–186. [CrossRef]
30. Backert, S.; Nielsen, B.L.; Borner, T. The mystery of the rings: Structure and replication of mitochondrial genomes from higher plants. *Trends. Plant Sci.* **1997**, *2*, 477–483. [CrossRef]
31. Backert, S.; Borner, T. Phage T4-like intermediates of DNA replication and recombination in the mitochondria of the higher plant Chenopodium album (L.). *Curr. Genet.* **2000**, *37*, 304–314. [CrossRef]
32. Sloan, D.B.; Alverson, A.J.; Chuckalovcak, J.P.; Wu, M.; McCauley, D.E.; Palmer, J.D.; Taylor, D.R. Rapid Evolution of Enormous, Multichromosomal Genomes in Flowering Plant Mitochondria with Exceptionally High Mutation Rates. *PLoS Biol.* **2012**, *10*, e1001241. [CrossRef]
33. Jacobs, M.A.; Payne, S.R.; Bendich, A.J. Moving pictures and pulsed-field gel electrophoresis show only linear mitochondrial DNA molecules from yeasts with linear-mapping and circular-mapping mitochondrial genomes. *Curr. Genet.* **1996**, *30*, 3–11. [CrossRef] [PubMed]
34. Hammani, K.; Giege, P. RNA metabolism in plant mitochondria. *Trends. Plant Sci.* **2014**, *19*, 380–389. [CrossRef] [PubMed]
35. Barcode of Life. Available online: http://www.barcodinglife.org/ (accessed on 1 July 2020).
36. Hollingsworth, P.M.; Forrest, L.L.; Spouge, J.L.; Hajibabaei, M.; Ratnasingham, S.; van der Bank, M.; Chase, M.W.; Cowan, R.S.; Erickson, D.L.; Fazekas, A.J.; et al. A DNA barcode for land plants. *Proc. Natl. Acad. Sci. USA* **2009**, *106*, 12794–12797.
37. Wang, Z.; Du, S.; Dayanandan, S.; Wang, D.; Zeng, Y.; Zhang, J. Phylogeny reconstruction and hybrid analysis of *Populus*. (Salicaceae) based on nucleotide sequences of multiple single-copy nuclear genes and plastid fragments. *PLoS ONE* **2014**, *9*, e103645. [CrossRef]
38. Mader, M.; Pakull, B.; Blanc-Jolivet, C.; Paulini-Drewes, M.; Bouda, Z.H.; Degen, B.; Small, I.; Kersten, B. Complete Chloroplast Genome Sequences of Four Meliaceae Species and Comparative Analyses. *Int. J. Mol. Sci.* **2018**, *19*, 701. [CrossRef]

39. Fladung, M.; Schroeder, H.; Wehenkel, C.; Kersten, B. Differentiation of six *Eucalyptus* trees grown in Mexico by ITS and six chloroplast barcoding markers. *Silvae Genet.* **2015**, *64*, 121–130. [CrossRef]

40. Birky, C.W. Uniparental inheritance of mitochondrial and chloroplast genes-Mechanisms and evolution. *Proc. Natl. Acad. Sci. USA* **1995**, *92*, 11331–11338. [CrossRef]

41. Ahmad, N.; Nielsen, B.L. Plant Organelle DNA Maintenance. *Plants* **2020**, *9*, 683. [CrossRef]

42. Morley, S.A.; Ahmad, N.; Nielsen, B.L. Plant Organelle Genome Replication. *Plants* **2019**, *8*, 358. [CrossRef]

43. Oldenburg, D.J.; Kumar, R.A.; Bendich, A.J. The amount and integrity of mtDNA in maize decline with development. *Planta* **2013**, *237*, 603–617. [CrossRef]

44. Kersten, B.; Rampant, P.F.; Mader, M.; Le Paslier, M.C.; Bounon, R.; Berard, A.; Vettori, C.; Schroeder, H.; Leple, J.C.; Fladung, M. Genome Sequences of *Populus. tremula.* Chloroplast and Mitochondrion: Implications for Holistic Poplar Breeding. *PLoS ONE* **2016**, *11*, e0147209. [CrossRef] [PubMed]

45. Li, X.; Yang, Y.; Henry, R.J.; Rossetto, M.; Wang, Y.; Chen, S. Plant DNA barcoding: From gene to genome. *Biol. Rev. Camb. Philos. Soc.* **2015**, *90*, 157–166. [CrossRef] [PubMed]

46. Nevill, P.G.; Zhong, X.; Tonti-Filippini, J.; Byrne, M.; Hislop, M.; Thiele, K.; van Leeuwen, S.; Boykin, L.M.; Small, I. Large scale genome skimming from herbarium material for accurate plant identification and phylogenomics. *Plant Methods* **2020**, *16*, 1. [CrossRef] [PubMed]

47. Saarela, J.M.; Burke, S.V.; Wysocki, W.P.; Barrett, M.D.; Clark, L.G.; Craine, J.M.; Peterson, P.M.; Soreng, R.J.; Vorontsova, M.S.; Duvall, M.R. A 250 plastome phylogeny of the grass family (Poaceae): Topological support under different data partitions. *PeerJ* **2018**, *6*, e4299. [CrossRef] [PubMed]

48. Vaughn, J.N.; Chaluvadi, S.R.; Rangan, L.; Bennetzen, J.L. Whole Plastome Sequences from Five Ginger Species Facilitate Marker Development and Define Limits to Barcode Methodology. *PLoS ONE* **2014**, *9*, e108581. [CrossRef]

49. Kersten, B.; Voss, M.; Fladung, M. Development of mitochondrial SNP markers to differentiate *Populus* species. *Trees–Struct. Funct.* **2015**, *29*, 575–582. [CrossRef]

50. Koch, G.; Haag, V.; Heinz, I. Pflichtbewusstsein: EUTR und Holzwerkstoffe—Prüfungen am Thünen-Kompetenzzentrum Holzherkünfte. *MDF. Mag. Co.* **2015**, *2015*, 64–67.

51. EU. *European Timber Regulation Verordnung (EU) Nr. 995/2010 des Europäischen Parlaments und des Rates vom 20;* Oktober 2010 über die Verpflichtungen von Marktteilnehmern, die Holz und Holzerzeugnisse in Verkehr bringen; EU: Brussels, Belgium, 2010.

52. The IUCN Red List of Threatened Species. Available online: http://www.iucnredlist.org/ (accessed on 3 July 2020).

53. Knorn, J.; Kuemmerle, T.; Radeloff, V.C.; Keeton, W.S.; Gancz, V.; Biris, I.A.; Svoboda, M.; Griffiths, P.; Hagatis, A.; Hostert, P. Continued loss of temperate old-growth forests in the Romanian Carpathians despite an increasing protected area network. *Environ. Conserv.* **2013**, *40*, 182–193. [CrossRef]

54. Liesebach, H.; Eusemann, P.; Liesebach, M. Verwandtschaftsbeziehungen innerhalb von Prüfgliedern in Herkunftsversuchen-Beispiel Buche (*Fagus sylvatica* L.) [Sibship structure in samples from a provenance trial. A case study in beech (*Fagus sylvatica* L.)]. *Forstarchiv* **2015**, *86*, 174–182.

55. Greiner, S.; Lehwark, P.; Bock, R. OrganellarGenomeDRAW (OGDRAW) version 1.3.1: Expanded toolkit for the graphical visualization of organellar genomes. *Nucleic Acids Res.* **2019**, *47*, W59–W64. [CrossRef]

56. OrganellarGenomeDRAW. Available online: http://ogdraw.mpimp-golm.mpg.de/ (accessed on 3 July 2020).

57. Lloyd Evans, D.; Hlongwane, T.T.; Joshi, S.V.; Riano Pachon, D.M. The sugarcane mitochondrial genome: Assembly, phylogenetics and transcriptomics. *PeerJ* **2019**, *7*, e7558. [CrossRef] [PubMed]

58. Zascavage, R.R.; Hall, C.L.; Thorson, K.; Mahmoud, M.; Sedlazeck, F.J.; Planz, J.V. Approaches to Whole Mitochondrial Genome Sequencing on the Oxford Nanopore MinION. *Curr. Protoc. Hum. Genet.* **2019**, *104*, e94. [CrossRef] [PubMed]

59. Frommer, B.; Holtgräwe, D.; Hausmann, L.; Viehöver, P.; Huettel, B.; Töpfer, R.; Weisshaar, B. Genome sequences of both organelles of the grapevine rootstock cultivar 'Börner'. *Microbiol. Resour. Announc.* **2020**, *9*, e01471-19. [CrossRef] [PubMed]

60. Adams, K.L.; Daley, D.O.; Whelan, J.; Palmer, J.D. Genes for two mitochondrial ribosomal proteins in flowering plants are derived from their chloroplast or cytosolic counterparts. *Plant Cell* **2002**, *14*, 931–943. [CrossRef]

61. Rivarola, M.; Foster, J.T.; Chan, A.P.; Williams, A.L.; Rice, D.W.; Liu, X.; Melake-Berhan, A.; Huot Creasy, H.; Puiu, D.; Rosovitz, M.J.; et al. Castor bean organelle genome sequencing and worldwide genetic diversity analysis. *PLoS ONE* **2011**, *6*, e21743. [CrossRef]

62. Shearman, J.R.; Sangsrakru, D.; Ruang-Areerate, P.; Sonthirod, C.; Uthaipaisanwong, P.; Yoocha, T.; Poopear, S.; Theerawattanasuk, K.; Tragoonrung, S.; Tangphatsornruang, S. Assembly and analysis of a male sterile rubber tree mitochondrial genome reveals DNA rearrangement events and a novel transcript. *BMC Plant Biol.* **2014**, *14*, 45. [CrossRef]

63. Choi, C.; Liu, Z.; Adams, K.L. Evolutionary transfers of mitochondrial genes to the nucleus in the *Populus* lineage and coexpression of nuclear and mitochondrial *Sdh4* genes. *New Phytol.* **2006**, *172*, 429–439. [CrossRef]

64. Choi, I.S.; Schwarz, E.N.; Ruhlman, T.A.; Khiyami, M.A.; Sabir, J.S.M.; Hajarah, N.H.; Sabir, M.J.; Rabah, S.O.; Jansen, R.K. Fluctuations in Fabaceae mitochondrial genome size and content are both ancient and recent. *BMC Plant Biol.* **2019**, *19*, 448. [CrossRef]

65. Alverson, A.J.; Wei, X.X.; Rice, D.W.; Stern, D.B.; Barry, K.; Palmer, J.D. Insights into the evolution of mitochondrial genome size from complete sequences of *Citrullus lanatus* and *Cucurbita pepo* (Cucurbitaceae). *Mol. Biol. Evol.* **2010**, *27*, 1436–1448. [CrossRef]

66. Adams, K.L.; Rosenblueth, M.; Qiu, Y.L.; Palmer, J.D. Multiple losses and transfers to the nucleus of two mitochondrial succinate dehydrogenase genes during angiosperm evolution. *Genetics* **2001**, *158*, 1289–1300.

67. Small, I.D.; Schallenberg-Rudinger, M.; Takenaka, M.; Mireau, H.; Ostersetzer-Biran, O. Plant organellar RNA editing: What 30 years of research has revealed. *Plant J.* **2020**, *101*, 1040–1056. [CrossRef] [PubMed]

68. Edera, A.A.; Gandini, C.L.; Sanchez-Puerta, M.V. Towards a comprehensive picture of C-to-U RNA editing sites in angiosperm mitochondria. *Plant Mol. Biol.* **2018**, *97*, 215–231. [CrossRef]

69. Quinones, V.; Zanlungo, S.; Holuigue, L.; Litvak, S.; Jordana, X. The *Cox1* Initiation Codon Is Created by Rna Editing in Potato Mitochondria. *Plant Physiol.* **1995**, *108*, 1327–1328. [CrossRef] [PubMed]

70. Brenner, W.G.; Mader, M.; Muller, N.A.; Hoenicka, H.; Schroeder, H.; Zorn, I.; Fladung, M.; Kersten, B. High Level of Conservation of Mitochondrial RNA Editing Sites Among Four *Populus* Species. *G3-Genes Genom. Genet.* **2019**, *9*, 709–717. [CrossRef]

71. Hollingsworth, P.M.; Graham, S.W.; Little, D.P. Choosing and Using a Plant DNA Barcode. *PLoS ONE* **2011**, *6*, e19254. [CrossRef] [PubMed]

72. Yang, C.H.; Wu, K.C.; Chuang, L.Y.; Chang, H.W. Decision Tree Algorithm-Generated Single-Nucleotide Polymorphism Barcodes of *rbcL* Genes for 38 Brassicaceae Species Tagging. *Evol. Bioinform.* **2018**, *14*, 1176934318760856. [CrossRef] [PubMed]

73. Arulandhu, A.J.; Staats, M.; Hagelaar, R.; Voorhuijzen, M.M.; Prins, T.W.; Scholtens, I.; Costessi, A.; Duijsings, D.; Rechenmann, F.; Gaspar, F.B.; et al. Development and validation of a multi-locus DNA metabarcoding method to identify endangered species in complex samples. *Gigascience* **2017**, *6*, 1–18. [CrossRef]

74. Lee, O.R.; Kim, M.K.; Yang, D.C. Authentication of medicinal plants by SNP-based multiplex PCR. *Methods Mol. Biol.* **2012**, *862*, 135–147.

75. Brunner, I.; Brodbeck, S.; Buchler, U.; Sperisen, C. Molecular identification of fine roots of trees from the Alps: Reliable and fast DNA extraction and PCR-RFLP analyses of plastid DNA. *Mol. Ecol.* **2001**, *10*, 2079–2087. [CrossRef]

76. Ju, L.P.; Shih, H.C.; Chiang, Y.C. Microsatellite primers for the endangered beech tree, *Fagus hayatae* (Fagaceae). *Am. J. Bot.* **2012**, *99*, e453–e456. [CrossRef]

77. Reutimann, O.; Gugerli, F.; Rellstab, C. A species-discriminatory single-nucleotide polymorphism set reveals maintenance of species integrity in hybridizing European white oaks (*Quercus* spp.) despite high levels of admixture. *Ann. Bot.* **2020**, *125*, 663–676. [CrossRef]

78. Schroeder, H.; Cronn, R.; Yanbaev, Y.; Jennings, T.; Mader, M.; Degen, B.; Kersten, B. Development of Molecular Markers for Determining Continental Origin of Wood from White Oaks (*Quercus* L. sect. Quercus). *PLoS ONE* **2016**, *11*, e0158221. [CrossRef] [PubMed]

79. Pencakowski, B.M.; Tokarski, M.; Jonkisz, A.; Czosnykowska-Lukacka, M.; Lenard, E.; Malodobra-Mazur, M. DNA profiling of oaks (*Quercus* spp.). *Arch. Med. Sadowej. Kryminol.* **2018**, *68*, 1–9. [PubMed]

80. Coutinho, J.P.; Carvalho, A.; Martin, A.; Lima-Brito, J. Molecular characterization of Fagaceae species using inter-primer binding site (iPBS) markers. *Mol. Biol. Rep.* **2018**, *45*, 133–142. [CrossRef] [PubMed]

81. Parker, J.; Helmstetter, A.J.; Devey, D.; Wilkinson, T.; Papadopulos, A.S.T. Field-based species identification of closely-related plants using real-time nanopore sequencing. *Sci. Rep.* **2017**, *7*, 8345. [CrossRef]

82. Dumolin, S.; Demesure, B.; Petit, R.J. Inheritance of chloroplast and mitochondrial genomes in pedunculate oak investigated with an efficient PCR method. *Theor. Appl. Genet.* **1995**, *91*, 1253–1256. [CrossRef]

83. Schalamun, M.; Nagar, R.; Kainer, D.; Beavan, E.; Eccles, D.; Rathjen, J.P.; Lanfear, R.; Schwessinger, B. Harnessing the MinION: An example of how to establish long-read sequencing in a laboratory using challenging plant tissue from *Eucalyptus pauciflora*. *Mol. Ecol. Resour.* **2019**, *19*, 77–89. [CrossRef]

84. Bolger, A.M.; Lohse, M.; Usadel, B. Trimmomatic: A flexible trimmer for Illumina sequence data. *Bioinformatics* **2014**, *30*, 2114–2120. [CrossRef]

85. Grabherr, M.G.; Haas, B.J.; Yassour, M.; Levin, J.Z.; Thompson, D.A.; Amit, I.; Adiconis, X.; Fan, L.; Raychowdhury, R.; Zeng, Q.D.; et al. Full-length transcriptome assembly from RNA-Seq data without a reference genome. *Nat. Biotechnol.* **2011**, *29*, 644–652. [CrossRef]

86. Tillich, M.; Lehwark, P.; Pellizzer, T.; Ulbricht-Jones, E.S.; Fischer, A.; Bock, R.; Greiner, S. GeSeq-versatile and accurate annotation of organelle genomes. *Nucleic Acids Res.* **2017**, *45*, W6–W11. [CrossRef]

87. Sequin-A DNA Sequence Submission Tool. 2019. Available online: https://www.ncbi.nlm.nih.gov/Sequin/ (accessed on 3 July 2020).

88. Sloan, D.B.; Wu, Z.Q.; Sharbrough, J. Correction of Persistent Errors in *Arabidopsis*. Reference Mitochondrial Genomes. *Plant Cell* **2018**, *30*, 525–527. [CrossRef] [PubMed]

89. Jung, J.; Kim, J.I.; Yi, G. geneCo: A visualized comparative genomic method to analyze multiple genome structures. *Bioinformatics* **2019**, *35*, 5303–5305. [CrossRef] [PubMed]

90. geneCo. The Visualization Software for Genome Data. Available online: https://bigdata.dongguk.edu/geneCo/ (accessed on 3 September 2020).

91. Stajich, J.E.; Block, D.; Boulez, K.; Brenner, S.E.; Chervitz, S.A.; Dagdigian, C.; Fuellen, G.; Gilbert, J.G.R.; Korf, I.; Lapp, H.; et al. The bioperl toolkit: Perl modules for the life sciences. *Genome Res.* **2002**, *12*, 1611–1618. [CrossRef] [PubMed]

92. Loytynoja, A. Phylogeny-aware alignment with PRANK. *Methods Mol. Biol.* **2014**, *1079*, 155–170. [PubMed]

93. GeSeq-Annotation of Organellar Genomes. Available online: https://chlorobox.mpimp-golm.mpg.de/geseq.html (accessed on 3 September 2020).

Article

Cultivated Tomato (*Solanum lycopersicum* L.) Suffered a Severe Cytoplasmic Bottleneck during Domestication: Implications from Chloroplast Genomes

Rachele Tamburino [1,†], Lorenza Sannino [1,†], Donata Cafasso [2], Concita Cantarella [3], Luigi Orrù [4], Teodoro Cardi [3], Salvatore Cozzolino [2], Nunzio D'Agostino [3,5] and Nunzia Scotti [1,*]

[1] CNR-IBBR, National Research Council of Italy, Institute of Biosciences and BioResources, Via Università 133, 80055 Portici (NA), Italy; rachele.tamburino@ibbr.cnr.it (R.T.); lorenza.sannino@ibbr.cnr.it (L.S.)

[2] Department of Biology, University of Naples Federico II, Via Cinthia, 80126 Naples, Italy; cafasso@unina.it (D.C.); cozzolin@unina.it (S.C.)

[3] CREA Research Centre for Vegetable and Ornamental Crops, Via dei Cavalleggeri 25, 84098 Pontecagnano Faiano (SA), Italy; concita.cantarella@gmail.com (C.C.); teodoro.cardi@crea.gov.it (T.C.); nunzio.dagostino@unina.it (N.D.)

[4] CREA Research Centre for Genomics and Bioinformatics, via S. Protaso 302, 29017 Fiorenzuola d'Arda (PC), Italy; luigi.orru@crea.gov.it

[5] Department of Agricultural Sciences, University of Naples Federico II, Via Università 133, 80055 Portici (NA), Italy

* Correspondence: nunzia.scotti@ibbr.cnr.it; Tel.: +39-0812-53-9482

† These authors contributed equally.

Received: 9 September 2020; Accepted: 22 October 2020; Published: 26 October 2020

Abstract: In various crops, genetic bottlenecks occurring through domestication can limit crop resilience to biotic and abiotic stresses. In the present study, we investigated nucleotide diversity in tomato chloroplast genome through sequencing seven plastomes of cultivated accessions from the Campania region (Southern Italy) and two wild species among the closest (*Solanum pimpinellifolium*) and most distantly related (*S. neorickii*) species to cultivated tomatoes. Comparative analyses among the chloroplast genomes sequenced in this work and those available in GenBank allowed evaluating the variability of plastomes and defining phylogenetic relationships. A dramatic reduction in genetic diversity was detected in cultivated tomatoes, nonetheless, a few *de novo* mutations, which still differentiated the cultivated tomatoes from the closest wild relative *S. pimpinellifolium*, were detected and are potentially utilizable as diagnostic markers. Phylogenetic analyses confirmed that *S. pimpinellifolium* is the closest ancestor of all cultivated tomatoes. Local accessions all clustered together and were strictly related with other cultivated tomatoes (*S. lycopersicum* group). Noteworthy, *S. lycopersicum* var. *cerasiforme* resulted in a mixture of both cultivated and wild tomato genotypes since one of the two analyzed accessions clustered with cultivated tomato, whereas the other with *S. pimpinellifolium*. Overall, our results revealed a very reduced cytoplasmic variability in cultivated tomatoes and suggest the occurrence of a cytoplasmic bottleneck during their domestication.

Keywords: next-generation sequencing; *Solanum*; Italian landraces; plastome; molecular markers; phylogenetic analysis

1. Introduction

Domestication of crops was one of the most complex and dynamic processes in plant evolution driven by humans, as it changed the distribution and frequency of plant species on the planet.

Crop domestication, through natural or artificial selection, generally results in a reduction of genetic diversity and in the loss of many adaptive traits from wild relatives [1,2]. The analysis of the genetic diversity of wild relatives and cultivated crops provided insight into the geographic and temporal details of domestication, whilst its estimation may provide the basis for developing suitable strategies for crop improvement, conservation and sustainable use [1]. Over past decades, molecular methods have been used to assess genetic diversity and, more recently, high throughput DNA sequencing technologies gave a huge boost to the estimation of genetic and adaptive diversity in crops and model plants [3–6].

Tomato (*Solanum lycopersicum* L.) is one of the most consumed vegetables in the world and belongs to the Solanaceae family, which includes species with a considerable economic importance (e.g., potato, pepper, eggplant, tobacco, and petunia) [7]. Within this family, *Solanum* is the largest and probably the most economically important genus, including both potatoes and tomatoes [8,9]. The original place of tomato domestication is still debated, however it is very likely that it occurred independently in the Peruvianum and Mexican regions [7]. The cultivated tomato, *S. lycopersicum* is divided into two botanical varieties *S. lycopersicum* var. *cerasiforme* (i.e., cherry tomato) and *S. lycopersicum* var. *lycopersicum*. Cherry tomato is native to the Andean region, but it also occurs in the subtropical areas and grows either as a true wild or cultivated species. For several years, cherry tomato has been considered an evolutionary intermediate between *S. pimpinellifolium*, the closest wild ancestor, and the cultivated *S. lycopersicum*. Recently, genetic studies [10] found cherry tomatoes were a mixture of wild and cultivated forms that likely originated from *S. pimpinellifolium*.

S. lycopersicum var. *lycopersicum* derived from cherry tomato through a multiphases process of domestication [11,12]. In particular, Blanca et al. [11] assumed a predomestication in the Andean regions that resulted into a wide morphological diversity of cherry tomatoes; then these genotypes reached Mesoamerica where the true domestication occurred. Here, traditional tomato varieties were developed and spread by Spanish conquistadors in Spain and Italy and, then, in the rest of the World. Since the late 18th century a strong selection activities has taken place in Europe, giving rise to a wide collection of tomato landraces adapted to local cultivation practices and environmental conditions [13–16]. More recently, these landraces gained increasing attention because of the high quality of fruits, their extended shelf-life and tolerance to environmental stresses [17–19]. Accordingly, several studies focused on the genome-wide characterization of the nuclear genetic diversity of various landraces [14–16,20–22].

Although it has been widely demonstrated the potentiality of cytoplasmic markers to study crop evolution and assess cytoplasmic bottlenecks occurred during the domestication history of several crops (i.e., rice, barley, potato, maize, and wheat) [23–28], to date little attention has been given to the analysis of the chloroplast genome in tomato landraces. Furthermore, a deeper knowledge of tomato plastomes would allow a better understanding of nuclear and cytoplasmic genome coevolution, and favor phylogenetic/barcoding studies and novel biotechnological approaches for breeding purposes [29–31].

In this work, we reported the complete plastome sequences of seven Italian cultivated tomato accessions grown in the Campania region (Southern Italy) and two wild species, namely *S. pimpinellifolium* and *S. neorickii*. Among Italian tomato accessions we selected the "Corbarino" landrace (processed tomato) characterized by obovoid fruits and moderate shelf-life, and six accessions belonging to the "Vesuviano" landrace (long shelf-life) characterized by hearth-shaped fruits with a pronounced pointed apex. Although they have the same place of origin, analysis based on nuclear single nucleotide polymorphisms (SNPs) showed a different clustering for some of these accessions [22]. The selected wild species are among the phylogenetically closest and most distantly related species to cultivated tomatoes and belong to two different phyletic groups characterized by red/orange- or green-fruited species, respectively. In particular, we aim to estimate the nucleotide diversity of tomato plastomes, inferring phylogenetic relationships, shedding lights on *de novo* mutations likely associated with the domestication and on the potential cytoplasmic bottleneck occurred during such a process.

2. Results

2.1. Chloroplast Genome Size and Organization

Sequencing of the nine tomato genotypes produced from 6.2 M (ves2001) to over 11.5 M (pol) high quality paired-end reads.

Cultivated tomato accessions, namely cor, pds, pol, ves2001, vfr, and vpz, had exactly the same plastome size (155,435 bp), with the exception of pgl (only one bp shorter), whereas plastome size in wild species was slightly larger in *S. neorickii* 1 (155,515 bp) and smaller in *S. pimpinellifolium* 1 (155,420 bp; Table 1). All genotypes exhibited the typical quadripartite structure of angiosperms plastome, including a pair of inverted repeats (IRs) separated by a large single copy (LSC) and a small single copy (SSC) regions (Table 1).

Table 1. Plastome features of the sequenced tomato genotypes.

Code	Species	Cultivar/Accession	Size (Base Pairs)			
			Total	LSC	SSC	IR
Cor	*S. lycopersicum*	Corbarino	155435	85857	18364	25607
Pds	*S. lycopersicum*	PDS	155435	85857	18364	25607
Pgl	*S. lycopersicum*	Piennolo giallo	155434	85857	18363	25607
Pol	*S. lycopersicum*	Pollena	155435	85857	18364	25607
ves2001	*S. lycopersicum*	Vesuvio 2001	155435	85857	18364	25607
Vfr	*S. lycopersicum*	Vesuvio foglia riccia	155435	85857	18364	25607
Vpz	*S. lycopersicum*	Vesuviano pizzo	155435	85857	18364	25607
S. neorickii 1	*S. neorickii*	LA2133	155515	85918	18379	25609
S. pimpinellifolium 1	*S. pimpinellifolium*	LA0722	155420	85842	18362	25608

2.2. Genetic Variability and Phylogenetic Analyses

Comparative analyses were performed in order to identify patterns of nucleotide variability among the tomato plastomes (the nine genotypes sequenced in this work and the twelve genotypes retrieved from GenBank). An overview of the nucleotide variability was shown in Figure S1. A variable number of SNPs (from a minimum of 9 to a maximum of 290) was observed when cultivated and wild plastomes were compared with the reference genome IPA-6 (Figure 1). Particularly, in cultivated tomatoes the number of SNPs was markedly low (from 9 to 17 SNPs), with the notable exception of cer1 that differed for 74 SNPs from IPA-6, a difference comparable to that of wild *S. pimpinellifolium*. All local accessions showed identical plastome sequences, with the exception of cor that differed for one point mutation in the exon 2 of the *rpoC1* gene and pgl that was one bp shorter.

Considering the low variability detected, to verify whether the SNPs identified in cultivated genotypes were ancestral or *de novo* mutations likely evolved before or after the domestication process, a comparative analysis was performed on the investigated tomato genotypes clustered into five groups: (1) the *S. lycopersicum* var. *lycopersicum* tomato commercial varieties IPA-6 and M82; (2) the seven local accessions from Campania region; (3) the *S. lycopersicum* var. *cerasiforme* cer1 and cer2; (4) the *S. pimpinellifolium*, *S. pimpinellifolium* 1 and *S. pimpinellifolium* 2, and (5) the wild including *S. habrochaites*, *S. cheesmaniae* and *S. galapagense*, phylogenetically closer to cultivated tomato than other wild species (Figure 2).

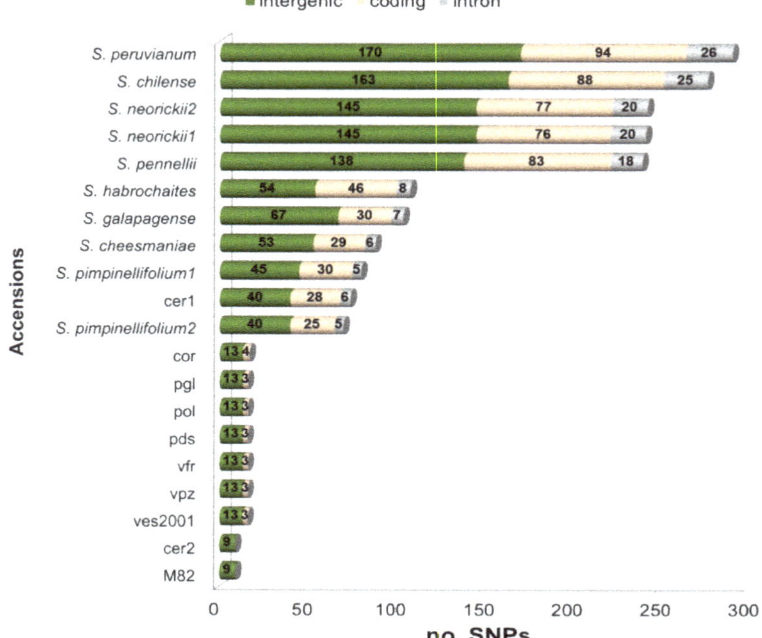

Figure 1. Stacked bar chart showing the distribution of single nucleotide polymorphisms (SNPs) that fall within coding sequences of genes, introns, and intergenic regions of the nine tomato plastomes sequenced in this work and in those of eleven species retrieved from GenBank. The plastome of IPA-6 (AM087200) was used as reference for SNP calling.

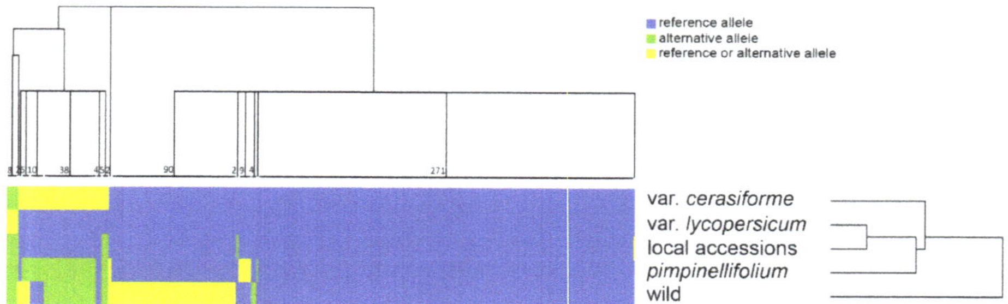

Figure 2. Hierarchical clustered heatmap representing color-coded SNP alleles as scored across 5 different groups of genotypes, i.e., var. *cerasiforme*; var. *lycopersicum*; local accessions; *pimpinellifolium*; wild species (including *S. habrochaites*, *S. cheesmaniae*, and *S. galapagense*). Numbers at the base of the tree indicate the SNP(s) that fall into each group. Blue: reference allele; green: alternative allele; yellow: reference or alternative allele.

Notably, a high number of SNPs (271) was common between all five groups and different from distantly related wild species, thus being ancestral mutations evolved in the phyletic lineages including cultivated tomatoes. Ninety SNPs were common between the cultivated tomatoes (*S. lycopersicum* var. *cerasiforme*, *S. lycopersicum* var. *lycopersicum* and local accessions) and the *S. pimpinellifolium* groups, whereas other wild species showed either the reference or the alternative allele. It is very likely that these latter 90 SNPs have been fixed only in the phyletic lineage of wild *S. pimpinellifolium* and cultivated tomatoes. Only two SNPs distinguished the local accessions from the remaining ones, but 38 SNPs (invariable among the cultivated genotypes) were different between cultivated tomatoes and *S. pimpinellifolium*. Notably, the *S. lycopersicum* var. *cerasiforme* group (cer1 and cer2) showed either the reference or the alternative allele of these 38 SNPs with cer1 sharing the *S. pimpinellifolium* allele, whilst cer2 the cultivated one. This result suggests that these 38 SNPs evolved as *de novo* mutations after the separation of cultivated forms from wild *S. pimpinellifolium* but were already present in the ancestral domesticated gene pool (including the *S. lycopersicum* var. *cerasiforme* group) and only subsequently fixed in the cultivated *S. lycopersicum* var. *lycopersicum* and local accessions groups. The other five SNPs were common between local accessions, *S. pimpinellifolium* and wild groups, while the *S. lycopersicum* var *lycopersicum* and the *S. lycopersicum* var. *cerasiforme* groups showed, respectively, the reference and either the reference or the alternative allele. By excluding cer1, in cultivated tomatoes seven SNPs (including the latter five and the two exclusive point mutations of local accessions) represent the only differences between plastomes of *S. lycopersicum* var. *lycopersicum* and local accessions.

As expected, wild species showed the highest number of SNPs independently from the phylogenetic distance to the reference genome (Figures 1 and 2) with variation detected even between accessions of the same species: ten different SNPs were found between the two accessions of *S. pimpinellifolium*. By looking at the distribution of SNPs in coding sequences, introns, and intergenic regions, the highest number of SNPs was scored in intergenic regions ranging from 9 to 13 in cultivated tomatoes and cer2, 40 in cer1, and 40-170 in the wild relatives. The same trend was observed for SNP distribution in coding sequences (Figure 1). Particularly, SNPs in wild species ranged from 25 to 94 and were dispersed as 1-2 variations per gene in most genes, whereas among cultivated genotypes up to four SNPs in local accessions were located in *matK*, exon 2 of the *rpoC1*, and *ycf1* coding sequences, one of these being in charge of an amino acid change. In contrast to all other cultivated landraces, once again, cer1 showed the number and distribution of SNPs similar to that found in the wild *S. pimpinellifolium*.

The most variable genes, especially among wild species, were *ndhH* and *ycf1* with 9 and 42 SNPs, respectively (Figure S2 and Figure S3). The mutations observed in the *ndhH* gene were synonymous (i.e., not causing changes in the amino acid sequence), whilst the nucleotide variability observed in *ycf1* was also reflected at the amino acid level. Interestingly, a SNP variation produced an amino acid change between the var. *lycopersicum* and the local accessions (Figure S3).

One hundred and fourteen simple sequence repeats (SSRs) were identified. The mononucleotide repeat (adenosine or thymine) was the most common type of microsatellite. Only four wild genotypes showed dinucleotide repeats (*S. neorickii* 1 and 2, *S. peruvianum*, and *S. chilense*). As observed for SNPs, clustered heatmap of SSRs across grouped genotypes revealed a very low level of polymorphism (Figure 3).

Sixty-seven SSRs were ancestral (same number of repeat units), being shared by all the analyzed genotypes as compared with the most distantly related wild species not included in the wild group; six SSRs have the same number of repeat units both in *S. lycopersicum* var. *lycopersicum* and local accessions groups, whereas *S. pimpinellifolium* and wild groups displayed a different number of repeat units, and *S. lycopersicum* var. *cerasiforme* both. One SSR displayed 13 repeats shared between *S. lycopersicum* var. *lycopersicum* and local accessions groups (i.e., *atpB-rbcL* intergenic region). An exclusive number of repeat units in the local accessions group was detected in the *ndhC-tRNA-Val* (UAC) intergenic region, while a number of repeat units exclusive of *S. lycopersicum* var. *lycopersicum* group was found in the *psbE-petL* intergenic region (Figure 3, Table S2). Interestingly, one SSR in the *atpH-atpI* intergenic region has the same number of repeat units both in all cultivated genotypes and wild group, while *S. pimpinellifolium* displayed a different number (Table S2). A complete description of SSR variability was shown in Figure S4a. As already observed for SNPs, SSRs were mainly located in intergenic regions (58%) and were mostly included in the LSC (75%; Figure S4b).

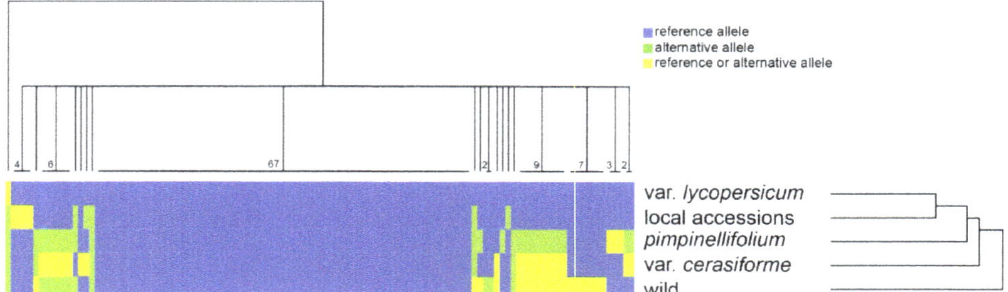

Figure 3. Hierarchical clustered heatmap representing color-coded simple sequence repeat (SSR) alleles as scored across 5 different groups of genotypes, i.e., var. *lycopersicum*; local accessions; *S. pimpinellifolium*; var. *cerasiforme*; wild species (including *S. habrochaites*, *S. cheesmaniae*, and *S. galapagense*). Numbers at the base of the tree indicate the SSR(s) that fall into each group. Blue: reference allele; green: alternative allele; yellow: reference or alternative allele.

Among the in silico identified microsatellites, eight SSR loci with small variation in the number of repeat units were experimentally tested to verify the correct estimation of their length. No variation in the number of repeat units was detected both in silico and in the electrophoresis profiles in a representation of the nine genotypes sequenced in this work and in a large dataset including additional local accessions and processed/fresh market tomatoes (e.g., Acampora, Lucariello, San Marzano, and Sorrento) confirming the absence of SSR variation within and among cultivated tomatoes. A notable exception was the one basis difference found in the microsatellite located in the *ndhF-rpl32* intergenic region that allowed distinguishing local accessions group from other tomato landraces and that was also confirmed by the electrophoresis profiles (data not shown).

Additionally, 17 perfect tandem repeats (TRs) were found, with cultivated species displaying a lower TR number when compared with wild species (Figure 4a). The identified TRs were mainly located in the LSC and intergenic regions (70 and 82%, respectively); two TRs found in all genotypes were in the coding region of the *rps16* and *rps4* genes (Figure 4b). The TR period size ranged from 13 to 26 bp (Figure 4c). TRs confirmed the low variability among the analyzed tomato genotypes. No TR was specific to any cultivated tomato; neither *de novo* TRs could be identified. A TR located in the *tRNA-Gln (UUG)-psbK* intergenic region was the only one to be found variable among species (Table S3). Particularly, local accessions and *S. pimpinellifolium* 1 had one copy, *S. neorickii* 1 and 2 had three copies, while *S. lycopersicum* var. *lycopersicum* (IPA-6 and M82), *S. lycopersicum* var. *cerasiforme* (cer1 and cer2), *S. pimpinellifolium* 2, and the remaining wild species had two copies (Table S3). Interestingly, a *de novo* duplication of four bases motif (ATAA)$_2$, exclusive of the local accessions, was scored by MSA (Figure S5).

Phylogenetic tree inferred from the complete plastomes of the twenty-one tomato genotypes using the potato chloroplast genome (*S. tuberosum* cv. Désirée, DQ386163) as an outgroup, showed two main clades with strong bootstrap support (100%; Figure 5). One clade included some wild species (*S. pennellii*, *S. neorickii* 1 and 2, *S. peruvianum*, and *S. chilense*) with *S. pennellii* as the basal species. The other clade is further separated into several subclades. In particular, the group that included the seven local accessions from the Campania region was closely related to a cluster populated by other cultivated varieties (IPA-6, M82, and cer2). As expected, all cultivated genotypes were more closely related to the clade comprising the two *S. pimpinellifolium* accessions and cer1. The remaining wild species (*S. galapagense*, *S. cheesmaniae*, and *S. habrochaites*) were in a separate clade. Finally, the phylogenetic analysis confirmed the admixed nature of *S. lycopersicum* var. *cerasiforme* as cer1 was closely related to the wild species (*S. pimpinellifolium* 1 and 2), while cer2 was part of the cultivated genotypes clade (M82 and IPA-6).

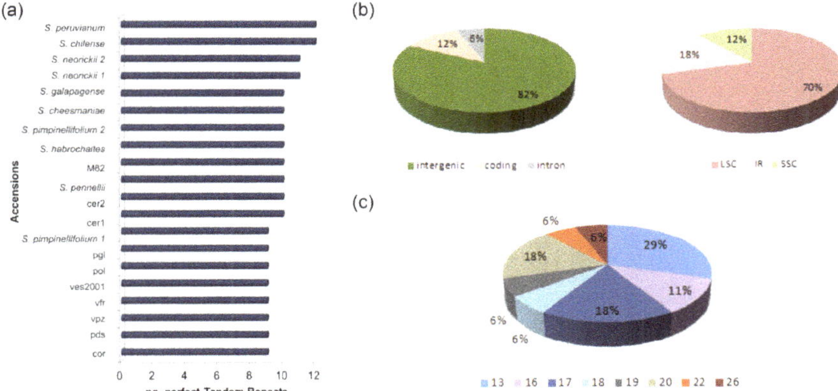

Figure 4. Perfect tandem repeats (TRs) in the nine plastomes sequenced in this work and in the plastome sequence of eleven species available in GenBank. The plastome of IPA-6 (AM087200) was used as reference. (**a**) Bar chart reporting the total number of TRs in each genotype. (**b**) Pie charts describing the percentage of TRs located in the coding sequences of genes, introns, and intergenic regions and in the large single copy (LSC), small single copy (SSC), and inverted repeat b (IR) regions. (**c**) Pie chart describing the percentage of TRs with a specific period size.

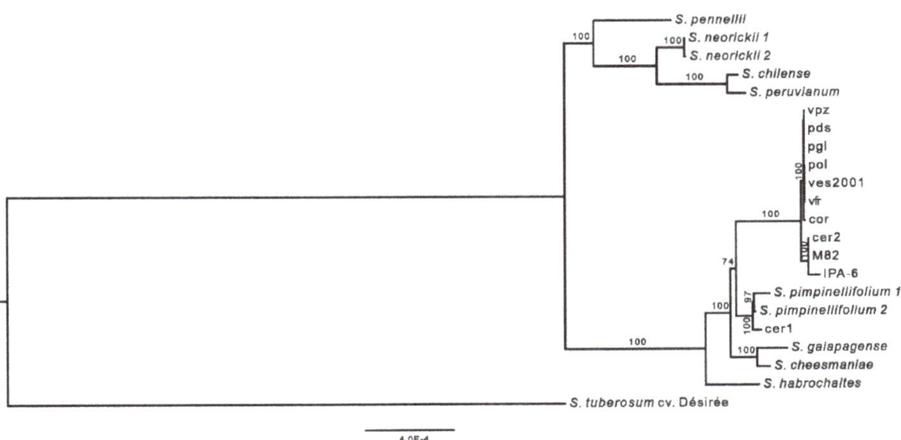

Figure 5. Phylogenetic tree of cultivated and wild tomato genotypes. Phylogram of the best maximum-likelihood (ML) tree on the complete plastome dataset using *Solanum tuberosum* cv. Désirée (DQ386163) as the outgroup. Numbers associated with branches are ML bootstrap support values. Bootstraps higher than 70% are reported on the nodes.

3. Discussion

Most crops experienced a reduction in genetic diversity (genetic bottleneck) due to the domestication process [32]. Indeed, the development of high yielding crops for food, feed, and other uses required the desirable phenotypes to be selected at the expense of variability present in their wild ancestors (founder effect) [33–35]. However, such "uniformity" often resulted in more vulnerable plants that are not able anymore to cope with biotic and abiotic stresses. As a consequence, wild relative species are often exploited as a reservoir of "exotic" alleles to secondarily increase variability in previously selected traits, thus favoring adaptation to changed conditions [34].

Landraces are locally adapted cultivars that are gaining increasing attention considering their typical traits (e.g., high quality of fruits and yield stability in low input agricultural systems) [17,19,36–38]. Although it has been widely demonstrated that the chloroplast genome is a valuable resource to study evolution and phylogenetic relationships among species [39,40], the genetic diversity of tomato landraces was largely based on the genome-wide characterization of their nuclear DNA variability [11,12,14–16,20–22,41]. Further, due the uniparental mode of inheritance, genetic bottleneck in organellar DNA may not necessarily reflect nuclear variability, thus providing additional/complementary information on the domestication process.

Comparative analyses of the nine plastomes sequenced in this work and of twelve plastomes retrieved from GenBank allowed both to evaluate the extent of the genetic bottleneck on the tomato chloroplast genome and define phylogenetic relationships among wild and cultivated accessions. For these aims, SNPs and SSRs were revealed to be more informative than TRs since no specific TR for cultivated tomato genotypes, or *de novo* TRs were identified in our survey.

Very low cpDNA variability was detected in tomato varieties with respect to that observed in wild species, thus indicating the occurrence of a very strong cytoplasmic bottleneck during domestication. The number of SNPs in wild species is 24-fold higher than in cultivated tomatoes (389 polymorphic SNPs out of 454 (86%)), while SSRs were slightly lower (49 polymorphic SSRs out of 114 (43%), 4-fold those observed in tomato varieties). The heterogeneous nature of the *S. lycopersicum* var. *cerasiforme* group is remarkable, namely, the two analyzed accessions showed a different behavior. Collected data and phylogeny clearly highlighted higher variability in cer1 compared with cer2 and suggest that although cer1 belongs to *cerasiforme* group, probably it was not subjected to the domestication process and can be considered as "wild" cultivated accession.

Detected levels of plastome variability are consistent with the extensive genetic erosion of cultivated tomato, especially in the light of the large diversity observed across wild relatives [5]. Similarly, pepper wild species displayed a number of SNP and SSR respectively 8-fold and 3-fold greater than that of cultivated genotypes [42].

Only 16 out of 454 SNPs were found polymorphic among cultivated tomato genotypes (3.5%). Comparable results were found in pepper varieties, where only the 4% of the scored SNP loci were polymorphic in cultivated accessions [42].

Similarly, only 12 out of 114 identified SSRs were polymorphic among cultivated tomato genotypes (11%). Comparable results were reported in cultivated barley showing one polymorphism out of seven analyzed SSRs (14%) [27] and pepper varieties, showing 19 polymorphic SSRs out of 92 (21%) [42]. Contrariwise, 16 out of 17 (94%) SSRs were polymorphic among cultivated bean [43].

As previously argued, genetic bottleneck at the nuclear level may not be reflected at the cytoplasmic level. An extreme cytoplasmic bottleneck has been previously hypothesized in cultivated potato by the analysis of SSR markers but no decreased levels of nuclear SSR diversity were recorded [26,39]. On the contrary, the genetic diversity analysis between American and European collections of common bean highlighted the absence of evident cytoplasmic bottleneck (only 2% loss of cpSSR diversity) [44], and a stronger nuclear bottleneck (30% loss of SSR diversity) [45] likely indicating that the founding common bean populations introduced in Europe were still highly variable in their cytoplasmic DNAs [46].

SNP arrays on some tomato cultivars, partially shared with this work (i.e., M82, cor, pgl, vfr, and ves2001), revealed a reduced nuclear genetic diversity [22].

Concordantly, the cpDNA analyses suggest an extreme low cytoplasmic variability of the founding cultivated tomato population. Indeed, cultivated varieties shared 361 out of 454 SNPs (79%) and 74 out of 114 SSRs (65%) with the ancestor *S. pimpinellifolium* (i.e., same SNP alleles and same SSR haplotypes) and only seven *de novo* SNPs and two *de novo* SSRs were different between *S. lycopersicum* var. *lycopersicum* and local accessions groups. All analyzed local accessions showed identical cpDNA sequences suggesting that these accessions have a unique domestication origin and that their cytoplasm has evolved monophyletically from the founder tomato gene pool, rather than representing an independent introduction. Still, the local accessions have distinctive sequences from the other commercial tomatoes (i.e., *S. lycopersicum* the var. *lycopersicum* group) excluding multiple independent selections of the obovoid fruits (Corbarino) or the hearth-shaped fruits with a pronounced pointed apex (the remaining accessions).

In this work, we also detected plastome variability between wild *S. pimpinellifolium* 1 and 2. These differences could be due to natural variability among accessions and/or possible errors in the sequencing/assembly procedure. The former hypotheses, however, is supported by differences also observed among other related wild species (*S. galapagense*, *S. cheesmaniae*, and *S. habrochaites*). Thus, the significant reduction in cpDNA variability found in the cultivated tomato gene pool can be directly ascribed as a consequence of the domestication process rather than to an already occurred loss of genetic variation in the closest wild relative, *S. pimpinellifolium*. Therefore, the present study suggests that a severe 'cytoplasmic bottleneck' occurred during the domestication of tomato, as has been reported in other crops: barley [27], lentil [47], onion [48], and potato [26].

A strict relationship between cultivated tomato varieties and the ancestor *S. pimpinellifolium* was supported by phylogeny. Species belonging to the Lycopersicon group (*S. lycopersicum*, *S. pimpinellifolium*, *S. cheesmaniae*, and *S. galapagense*) [49] form a well-supported clade in agreement with previous phylogenetic studies [5].

In particular, all local accessions clustered together in a subgroup with *S. lycopersicum* var. *lycopersicum* and cer 2. On the contrary, some accessions (i.e., cor, pgl, vfr, and ves2001) were grouped in different clusters based on nuclear SNP genotyping [22]. Noteworthy, cer1 was included in the same group of *S. pimpinellifolium* accessions, thus plastome diversity analysis confirmed the mixed nature of *S. lycopersicum* var. *cerasiforme* as previously observed with the analysis of nuclear variability [10,11,50,51].

The observed low variability of the cultivated tomatoes chloroplast genome can be explained by taking into account both the genetic bottleneck during their domestication and its low mutation rate. Notably, comparison of the plastome sequences of the two modern tomato varieties IPA-6 and Ailsa Craig, the former bred in South America and the latter in Europe, resulted in identical cpDNA sequences, thus demonstrating the stability of plastome in tomato cultivars over a period of at least a few hundred years of separation [52] without the insurgence of any *de novo* mutation. Although low variation is the rule in tomato cpDNA, few plastid regions have been identified that might be exploited as diagnostic markers: two *de novo* SNPs, one SSR and a short sequence duplication (ATAA)$_2$ were exclusive of all local accessions, whereas, one SSR was typical of all the var. *lycopersicum* group.

Variability found in all tomato genotypes mainly affected intergenic regions. However, the most variable genes were *ycf1* (showing both synonymous and non-synonymous mutations) and *ndhH*. Both these genes have been proposed as tools to resolve the phylogenetic relationships among closely related genera and species [53–55] and at least *ycf1* was found variable even within cultivated plastomes leading to amino acid change (Figure S3).

Overall, our work contributes to the characterization of tomato plastid genomes and their phylogenetic relationships, and especially highlights the severe reduction in variability at plastid DNA as a consequence of the strong genetic bottleneck occurred in the founding population during the domestication process.

4. Materials and Methods

4.1. Plant Material

Seven Italian cultivated tomato accessions grown in the Campania region (Southern Italy), Corbarino (cor) landrace, and six accessions belonging to the "Vesuviano" landrace, Pollena (pol), PDS (pds), Vesuvio 2001 (ves2001), Vesuviano foglia riccia (vfr), Vesuviano pizzo (vpz), Piennolo giallo (pgl), and two wild tomato species, *S. pimpinellifolium* (LA0722, Peru) and *S. neorickii* (LA2133, Ecuador) were sampled for chloroplast isolation, DNA extraction and sequencing. Drs. M.S. Grillo and S. Grandillo from the CNR, Institute of Bioscience and BioResources, Portici, kindly provided the seeds.

4.2. Chloroplast DNA Isolation and Extraction

Plants were kept in the dark for 48 h before harvesting to reduce starch accumulation. Fresh leaves (15–25 g) were collected and used for chloroplast isolation with discontinuous sucrose gradient [56]. Purified chloroplasts were lysed with a detergent and proteins eliminated by proteinase K and phenol/chloroform treatments following the procedure described by [57].

4.3. Genome Sequencing, Assembling, and Annotation

DNA samples were sequenced using the GA II Illumina sequencer (2 × 75 paired-end reads with an estimated inset size of 400 bp). Quality check on raw reads was performed using FastQC v.0.11.2 (http://www.bioinformatics.babraham.ac.uk/projects/fastqc/). Then, the fastq_quality_filter utility from the FASTX-toolkit (http://hannonlab.cshl.edu/fastx_toolkit/) was used to remove sequences with a quality score equal or lower than 30 in more than 90% of the read length. Illumina technical sequences were removed by using Trimmomatic v.0.32 [58]. Reference-based assembly was performed using the Columbus module within the Velvet package [59] with a k-mer size of 65. The chloroplast genome sequence of *S. lycopersicum* cv. IPA-6 (AM087200) was used as reference. Contigs were ordered and oriented by using ABACAS [60] for the final assembly. Finally, high quality reads were aligned back onto the assemblies using Bowtie2 [61] with default settings to validate and manually fix errors in the assemblies. Per base genome coverage was computed using the genomecov utility of bedtools version 2.20.1 (Figure S6) [62]. The annotation of chloroplast genomes was performed using GeSeq (https://chlorobox.mpimp-golm.mpg.de/geseq.html). Gene annotations were manually curated using *S. lycopersicum* cv. IPA-6 (AM087200) annotations as reference. Chloroplast genome sequences and annotations produced in this study can be found in GenBank under accession numbers MT811790-MT811798.

4.4. Detection and Analysis of Sequence Variations

Single nucleotide variants (SNVs) were identified using the snp-sites tool (https://github.com/sanger-pathogens/snp-sites). Such a tool extracts SNPs from a multiple sequence alignment using the cpDNA of *S. lycopersicum* cv. IPA-6 as reference sequence. SNP annotation was manually curated.

The microsatellite (MISA) identification tool (http://pgrc.ipk-gatersleben.de/misa/) was run to identify microsatellites (SSR) using the unit_size/min_repeats parameters as follows: 1/8, 2/6, 3/5, 4/5, 5/5, 6/5. The Tandem Repeat Finder web tool accessible at https://tandem.bu.edu/trf/trf.basic.submit.html was used to detect perfect tandem repeats with default settings.

In silico identified SSR loci were experimentally tested for variation in the number of repeat units. For this aim, 8 SSR loci were selected from the MISA output by focusing on those with small variation in the number of repeat units to verify the correct estimation of their repeat length. Primers were designed with Primer3 (http://frodo.wi.mit.edu/primer3/). The primer size was set from 18 to 25 bp, the Tm ranged from 51 to 59 °C and the other parameters were set as default (Table S1). For each microsatellite locus, the forward primers were labeled with the different fluorescent dyes 6-FAM, ATTO550, ATTO565, and HEX (Sigma Aldrich, USA). Beside the sequenced local accessions, we applied these primers to 19 additional local genotypes, namely further seven local accessions and twelve processed/fresh market tomatoes.

All PCR amplifications were performed by a Perkin Elmer 9700 thermocycler according to PCR conditions as reported in [63]. The conditions were maintained constant for all loci in order to maximize standardization. Amplified microsatellite products were then genotyped using an Applied Biosystem 3130 automatic sequencer with LIZ (500) as an internal standard and sized with GENEMAPPER software v. 3.7 (Thermo Fisher Scientific-Applera, USA).

Multiple sequence alignments (MSA) were generated using MAFFT version 7 [64] with default settings. Single-nucleotide variants were identified by the snp-sites software [65] using as input the plastomes MSA and the cpDNA of *S. lycopersicum* cv. IPA-6 (AM087200) as reference. To highlight differences among nucleotide sequences of plastomes, MSA were visualized using the NCBI Multiple Sequence Alignment Viewer available at https://www.ncbi.nlm.nih.gov/projects/msaviewer/.

RAxML [66] was used to build a maximum-likelihood (ML) tree with 10,000 rapid bootstrap inferences, a generalized time reversible (GTR) substitution matrix and Gamma model of rate heterogeneity. The plastome of *S. tuberosum* cv. Désirée (DQ386163) was used as the outgroup. The ML tree was visualized with FigTree v.1.4.2 (http://tree.bio.ed.ac.uk/software/figtree/).

In addition to *S. lycopersicum* cv. IPA-6 (AM087200), eleven tomato genotypes available in GenBank: *S. peruvianum* (KP117026), *S. chilense* (KP117021), *S. neorickii* (*S. neorickii* 2, KP117025), *S. pennellii* (HG975452), *S. habrochaites* (KP117023), *S. galapagense* (NC_026878), *S. cheesmaniae* (NC_026876), *S. pimpinellifolium* (*S. pimpinellifolium* 2, KP117027), *S. lycopersicum* (cv M82, HG975525), and *S. lycopersicum* var. *cerasiforme* (cer1, KY887588; and cer2, KY887587) were retrieved for comparative analyses. Heatmaps were generated using Morpheus (https://software.broadinstitute.org/morpheus). Single-linkage hierarchical clustering on both rows and columns was based on the metric "Euclidean distance".

Supplementary Materials: The following are available online at http://www.mdpi.com/2223-7747/9/11/1443/s1, Figure S1: Overview of the nucleotide variability in nine plastomes sequenced in this study and in eleven species available in GenBank. The accession number AM087200 (cv. IPA-6) was used as reference. Red lines represent variable regions, Figure S2: Schematic representation of the nucleotide variability observed in the *ndhH* gene for the plastomes under investigation. Grey bar represents the nucleotide multiple-sequence alignment (MSA) and it is scaled according to the MSA length. Black boxes indicate variable regions in the MSA. Above and below each box, a snapshot of the MSA along with alignment positions is reported, Figure S3: Schematic representation of the amino acid variability observed in the Ycf1 protein for the plastomes under investigation. Grey bar represents the amino acid multiple-sequence alignment (MSA) and it is scaled according to the MSA length. Black boxes indicate variable regions in the MSA. Above and below each box, a snapshot of the MSA along with alignment positions is reported; Figure S4: Simple sequence repeats (SSRs) in nine plastomes sequenced in this study and in eleven species available in GenBank. The plastome of IPA-6 (AM087200) was used as reference a) Heatmap representing differences in SSR size; colors range from red (SSR size larger than the reference) through yellow to blue (SSR size smaller than the reference). Black is for missing SSRs. b) Pie chart describing the percentage of SSRs located in coding sequences of genes, introns, and intergenic regions and in the large single copy (LSC), small single copy (SSC), and inverted repeat b (IR) regions, Figure S5: Multiple-sequence alignment (MSA) of the region harboring the duplicated sequence (ATAA)$_2$ scored only in local landraces, Figure S6: Distribution of per-base sequencing depth for each chloroplast genome sequenced in this work. The average coverage per-base is also reported. Table S1: Tomato cpSSR primers developed in this study, Table S2: Simple sequence repeats (SSRs) in the twenty-one tomato chloroplast genomes using IPA-6 (AM087200) as the reference genome. SSRs size, location, and distribution among different regions, namely coding, intron, and intergenic are reported. The unit_size/min_repeats parameters were as follows: 1/8, 2/6, 3/5, 4/5, 5/5, and 6/5. SSRs located in IRa were not counted. SSRs were identified using MISA—microsatellite identification tool (http://pgrc.ipk-gatersleben.de/misa/), Table S3: Tandem repeats (TRs) in the twenty-one tomato chloroplast genomes using IPA-6 (AM087200) as the reference genome. TRs copy number and location are reported. TRs were identified using the tool available at https://tandem.bu.edu/trf/trf.basic.submit.html.

Author Contributions: T.C., S.C., N.D. and N.S. conceived and designed the research. C.C., N.D. performed bioinformatic analyses. R.T., L.S., D.C. and N.S. carried out wet-lab experiments. L.O. performed Illumina sequencing of plastomes. R.T., D.C., S.C., N.D. and N.S. contributed to data interpretation. R.T. and N.S. wrote the manuscript. T.C., S.C., N.D. and N.S. revised the manuscript. All authors have read and agreed to the published version of the manuscript.

Funding: This research was partially funded by grant from the Italian Ministry of Economy and Finance (MEF)-National Research Council of Italy (CNR), Project "Conoscenze Integrate per la Sostenibilità e l'Innovazione del *made in Italy* Agroalimentare" (CISIA), Legge n. 191/2009.

Acknowledgments: Technical assistance of Mr. G. Guarino and Mr. R. Nocerino (CNR-IBBR, Portici, Italy) with artworks and plant growth is gratefully acknowledged.

Conflicts of Interest: The authors declare no conflict of interest.

References

1. Meyer, R.S.; DuVal, A.E.; Jensen, H.R. Patterns and processes in crop domestication: An historical review and quantitative analysis of 203 global food crops. *New Phytol.* **2012**, *196*, 29–48. [CrossRef] [PubMed]

2. Smykal, P.; Nelson, M.N.; Berger, J.D.; von Wettberg, E.J.B. The impact of genetic changes during crop domestication. *Agronomy* **2018**, *8*, 119. [CrossRef]

3. Weigel, D.; Mott, R. The 1001 genomes project for *Arabidopsis thaliana*. *Genome Biol.* **2009**, *10*, 107. [CrossRef] [PubMed]

4. The Tomato Genome Consortium. The tomato genome sequence provides insights into fleshy fruit evolution. *Nature* **2012**, *485*, 635–641. [CrossRef]

5. Aflitos, S.; Schijlen, E.; de Jong, H.; de Ridder, D.; Smit, S.; Finkers, R.; Wang, J.; Zhang, G.; Li, N.; Mao, L.; et al. Exploring genetic variation in the tomato (*Solanum* section *Lycopersicon*) clade by whole-genome sequencing. *Plant J.* **2014**, *80*, 136–148. [CrossRef]

6. Acquadro, A.; Barchi, L.; Portis, E.; Nourdine, M.; Carli, C.; Monge, S.; Valentino, D.; Lanteri, S. Whole genome resequencing of four Italian sweet pepper landraces provides insights on sequence variation in genes of agronomic value. *Sci. Rep.* **2020**, *10*, 9189. [CrossRef]

7. Bergougnoux, V. The history of tomato: From domestication to biopharming. *Biotechnol. Adv.* **2014**, *32*, 170–189. [CrossRef]

8. Peralta, I.E.; Spooner, D.M. Granule-bound starch synthase (GBSSI) gene phylogeny of wild tomatoes (*Solanum* L. section *Lycopersicon* [Mill.] Wettst. subsection *Lycopersicon*). *Am. J. Bot.* **2001**, *88*, 1888–1902. [CrossRef]

9. Peralta, I.E.; Spooner, D.M. *History, Origin and Early Cultivation of Tomato (Solanaceae)*; Razdan, M.K., Matto, A.K., Eds.; Science Publishers: Plymouth, MA, USA, 2007; Volume Tomato: Genetic Improvement of Solanaceous Crops; pp. 1–27.

10. Nesbitt, T.C.; Tanksley, S.D. Comparative sequencing in the genus *Lycopersicon*: Implication for the evolution of fruit size in the domestication of the cultivated tomatoes. *Genetics* **2002**, *162*, 365–379.

11. Blanca, J.; Canizares, J.; Cordero, L.; Pascual, L.; Diez, M.J.; Nuez, F. Variation revealed by SNP genotyping and morphology provides insight into the origin of the tomato. *PLoS ONE* **2012**, *7*, e48198. [CrossRef]

12. Blanca, J.; Montero-Pau, J.; Sauvage, C.; Bauchet, G.; Illa, E.; Diez, M.J.; Francis, D.; Causse, M.; van der Knaap, E.; Canizares, J. Genomic variation in tomato, from wild ancestors to contemporary breeding accessions. *BMC Genom.* **2015**, *16*, 257. [CrossRef] [PubMed]

13. Bai, Y.; Lindhout, P. Domestication and breeding of tomatoes: What have we gained and what can we gain in the future? *Ann. Bot.* **2007**, *100*, 1085–1094. [CrossRef]

14. Corrado, G.; Caramante, M.; Piffanelli, P.; Rao, R. Genetic diversity in Italian tomato landraces: Implications for the development of a core collection. *Sci. Hortic.* **2014**, *168*, 138–144. [CrossRef]

15. Corrado, G.; Piffanelli, P.; Caramante, M.; Coppola, M.; Rao, R. SNP genotyping reveals genetic diversity between cultivated landraces and contemporary varieties of tomato. *BMC Genom.* **2013**, *14*, 835. [CrossRef] [PubMed]

16. Ercolano, M.R.; Sacco, A.; Ferriello, F.; D'Alessandro, R.; Tononi, P.; Traini, A.; Barone, A.; Zago, E.; Chiusano, M.L.; Buson, G.; et al. Patchwork sequencing of tomato San Marzano and Vesuviano varieties highlights genome-wide variations. *BMC Genom.* **2014**, *15*, 138. [CrossRef]

17. Andreakis, N.; Giordano, I.; Pentangelo, A.; Fogliano, V.; Graziani, G.; Monti, L.M.; Rao, R. DNA fingerprinting and quality traits of Corbarino cherry-like tomato landraces. *J. Agric. Food Chem.* **2004**, *52*, 3366–3371. [CrossRef]

18. Digilio, M.C.; Corrado, G.; Sasso, R.; Coppola, V.; Iodice, L.; Pasquariello, M.; Bossi, S.; Maffei, M.E.; Coppola, M.; Pennacchio, F.; et al. Molecular and chemical mechanisms involved in aphid resistance in cultivated tomato. *New Phytol.* **2010**, *187*, 1089–1101. [CrossRef]

19. Galmés, J.; Conesa, M.À.; Ochogavía, J.M.; Perdomo, J.A.; Francis, D.M.; Ribas, M.; Savé, R.; Flexas, J.; Medrano, H.; Cifre, J. Physiological and morphological adaptations in relation to water use efficiency in Mediterranean accessions of *Solanum lycopersicum*. *Plant Cell Environ.* **2011**, *34*, 245–260. [CrossRef]

20. Sacco, A.; Ruggieri, V.; Parisi, M.; Festa, G.; Rigano, M.M.; Picarella, M.E.; Mazzucato, A.; Barone, A. Exploring a tomato landraces collection for fruit-related traits by the aid of a high-throughput genomic platform. *PLoS ONE* **2015**, *10*, e0137139. [CrossRef] [PubMed]

21. Tranchida-Lombardo, V.; Aiese Cigliano, R.; Anzar, I.; Landi, S.; Palombieri, S.; Colantuono, C.; Bostan, H.; Termolino, P.; Aversano, R.; Batelli, G.; et al. Whole-genome re-sequencing of two Italian tomato landraces reveals sequence variations in genes associated with stress tolerance, fruit quality and long shelf-life traits. *DNA Res.* **2018**, *25*, 149–160. [CrossRef]

22. Tranchida-Lombardo, V.; Mercati, F.; Avino, M.; Punzo, P.; Fiore, M.C.; Poma, I.; Patanè, C.; Guarracino, M.R.; Sunseri, F.; Tucci, M.; et al. Genetic diversity in a collection of Italian long storage tomato landraces as revealed by SNP markers array. *Plant Biosyst.* **2019**, *153*, 288–297. [CrossRef]
23. Provan, J.; Corbett, G.; McNicol, J.W.; Powell, W. Chloroplast DNA variability in wild and cultivated rice (*Oryza* spp.) revealed by polymorphic chloroplast simple sequence repeats. *Genome* **1997**, *40*, 104–110. [CrossRef] [PubMed]
24. Provan, J.; Corbett, G.; Waugh, R.; McNicol, J.W.; Morgante, M.; Powell, W. DNA fingerprints of rice (*Oryza sativa*) obtained from hypervariable chloroplast simple sequence repeats. *Proc. R. Soc. Lond. Ser. B Biol. Sci.* **1996**, *263*, 1275–1281. [CrossRef]
25. Provan, J.; Lawrenc, P.; Young, G.; Wright, F.; Bird, R.; Paglia, G.P.; Cattonaro, F.; Morgante, M.; Powell, W. Analysis of the genus *Zea* (Poaceae) using polymorphic chloroplast simple sequence repeats. *Plant Syst. Evol.* **1999**, *218*, 245–256. [CrossRef]
26. Provan, J.; Powell, W.; Dewar, H.; Bryan, G.; Machray, G.C.; Waugh, R. An extreme cytoplasmic bottleneck in the modern European cultivated potato (*Solanum tuberosum*) is not reflected in decreased levels of nuclear diversity. *Proc. R. Soc. Lond. Ser. B Biol. Sci.* **1999**, *266*, 633–639. [CrossRef]
27. Provan, J.; Russell, J.R.; Booth, A.; Powell, W. Polymorphic chloroplast simple sequence repeat primers for systematic and population studies in the genus *Hordeum*. *Mol. Ecol.* **1999**, *8*, 505–511. [CrossRef]
28. Provan, J.; Wolters, P.; Caldwell, K.H.; Powell, W. High-resolution organellar genome analysis of *Triticum* and *Aegilops* sheds new light on cytoplasm evolution in wheat. *Theor. Appl. Genet.* **2004**, *108*, 1182–1190. [CrossRef]
29. Rogalski, M.; do Nascimento Vieira, L.; Fraga, H.P.; Guerra, M.P. Plastid genomics in horticultural species: Importance and applications for plant population genetics, evolution, and biotechnology. *Front. Plant Sci.* **2015**, *6*, 586. [CrossRef]
30. Daniell, H.; Lin, C.S.; Yu, M.; Chang, W.J. Chloroplast genomes: Diversity, evolution, and applications in genetic engineering. *Genome Biol.* **2016**, *17*, 134. [CrossRef]
31. Tonti-Filippini, J.; Nevill, P.G.; Dixon, K.; Small, I. What can we do with 1000 plastid genomes? *Plant J.* **2017**, *90*, 808–818. [CrossRef]
32. Doebley, J.F.; Gaut, B.S.; Smith, B.D. The molecular genetics of crop domestication. *Cell* **2006**, *127*, 1309–1321. [CrossRef]
33. Ladizinsky, G. Founder effect in crop-plant evolution. *Econ. Bot.* **1985**, *39*, 191–199. [CrossRef]
34. Tanksley, S.D.; McCouch, S.R. Seed banks and molecular maps: Unlocking genetic potential from the wild. *Science* **1997**, *277*, 1063–1066. [CrossRef]
35. Saini, P.; Saini, P.; Kaur, J.J.; Francies, R.M.; Gani, M.; Rajendra, A.A.; Negi, N.; Jagtap, A.; Kadam, A.; Singh, C.; et al. Molecular approaches for harvesting natural diversity for crop improvement. In *Rediscovery of Genetic and Genomic Resources for Future Food Security*; Salgotra, R., Zargar, S., Eds.; Springer: Singapore, 2020; pp. 67–169.
36. Fernie, A.R.; Tadmor, Y.; Zamir, D. Natural genetic variation for improving crop quality. *Curr. Opin. Plant Biol.* **2006**, *9*, 196–202. [CrossRef]
37. Berg, T. Landraces and folk varieties: A conceptual reappraisal of terminology. *Euphytica* **2009**, *166*, 423–430. [CrossRef]
38. Conesa, M.A.; Fullana-Pericas, M.; Granell, A.; Galmes, J. Mediterranean long shelf-life landraces: An untapped genetic resource for tomato improvement. *Front. Plant Sci.* **2020**, *10*, 1651. [CrossRef]
39. Provan, J.; Powell, W.; Hollingsworth, P.M. Chloroplast microsatellites: New tools for studies in plant ecology and evolution. *Trends Ecol. Evol.* **2001**, *16*, 142–147. [CrossRef]
40. Song, Y.; Wang, S.; Ding, Y.; Xu, J.; Li, M.F.; Zhu, S.; Chen, N. Chloroplast genomic resource of *Paris* for species discrimination. *Sci. Rep.* **2017**, *7*, 3427. [CrossRef]
41. Sim, S.C.; Robbins, M.D.; Van Deynze, A.; Michel, A.P.; Francis, D.M. Population structure and genetic differentiation associated with breeding history and selection in tomato (*Solanum lycopersicum* L.). *Heredity* **2011**, *106*, 927–935. [CrossRef]
42. D'Agostino, N.; Tamburino, R.; Cantarella, C.; De Carluccio, V.; Sannino, L.; Cozzolino, S.; Cardi, T.; Scotti, N. The complete plastome sequences of eleven *Capsicum* genotypes: Insights into DNA variation and molecular evolution. *Genes* **2018**, *9*, 503. [CrossRef]

43. Desiderio, F.; Bitocchi, E.; Bellucci, E.; Rau, D.; Rodriguez, M.; Attene, G.; Papa, R.; Nanni, L. Chloroplast microsatellite diversity in *Phaseolus vulgaris*. *Front. Plant Sci.* **2013**, *3*, 312. [CrossRef]

44. Angioi, S.A.; Rau, D.; Attene, G.; Nanni, L.; Bellucci, E.; Logozzo, G.; Negri, V.; Zeuli, P.L.S.; Papa, R. Beans in Europe: Origin and structure of the European landraces of *Phaseolus vulgaris* L. *Theor. Appl. Genet.* **2010**, *121*, 829–843. [CrossRef]

45. Papa, R.; Nanni, L.; Sicard, D.; Rao, D.; Attene, G. The evolution of genetic diversity in *Phaseolus vulgaris*. In *Darwin's Harvest: New Approaches to the Origins, Evolution and Conservation of Crops*; Motley, T.J., Zerega, N., Cross, H., Eds.; Columbia University Press: NewYork, NY, USA, 2006; pp. 121–142.

46. Bellucci, E.; Bitocchi, E.; Rau, D.; Rodriguez, M.; Biagetti, E.; Giardini, A.; Attene, G.; Nanni, L.; Papa, R. Genomics of origin, domestication and evolution of *Phaseolus vulgaris*. In *Genomics of Plant Genetic Resources*; Tuberosa, R., Graner, A., Frison, E., Eds.; Springer: Singapore, 2014; Volume 1, pp. 483–507.

47. Ladizinsky, G. Identification of the lentil wild genetic stock. *Genet. Resour. Crop Evol.* **1999**, *46*, 115–118. [CrossRef]

48. Friesen, N.; Pollner, S.; Bachmann, K.; Blattner, F.R. RAPDs and noncoding chloroplast DNA reveal a single origin of the cultivated *Allium fistulosum* from *A. altaicum* (Alliaceae). *Am. J. Bot.* **1999**, *86*, 554–562. [CrossRef]

49. Peralta, I.E.; Spooner, D.M.; Knapp, S. *Taxonomy of Wild Tomatoes and Their Relatives (Solanum sect. Lycopersicoides, sect Jugandifolia, sect. Lycopersicon; Solanaceae)*; Anderson, C., Ed.; American Society of Plant Taxonomists: Ann Arbor, MI, USA, 2008; pp. 1–186.

50. Ranc, N.; Muños, S.; Xu, J.; Le Paslier, M.C.; Chauveau, A.; Bounon, R.; Rolland, S.; Bouchet, J.P.; Brunel, D.; Causse, M. Genome-wide association mapping in tomato (*Solanum lycopersicum*) is possible using genome admixture of *Solanum lycopersicum* var. *cerasiforme*. *G3* **2012**, *2*, 853–864. [CrossRef] [PubMed]

51. Causse, M.; Desplat, N.; Pascual, L.; Le Paslier, M.-C.; Sauvage, C.; Bauchet, G.; Bérard, A.; Bounon, R.; Tchoumakov, M.; Brunel, D.; et al. Whole genome resequencing in tomato reveals variation associated with introgression and breeding events. *BMC Genom.* **2013**, *14*, 791. [CrossRef]

52. Kahlau, S.; Aspinall, S.; Gray, J.C.; Bock, R. Sequence of the tomato chloroplast DNA and evolutionary comparison of solanaceous plastid genomes. *J. Mol. Evol.* **2006**, *63*, 194–207. [CrossRef]

53. Martín, M.; Sabater, B. Plastid *ndh* genes in plant evolution. *Plant Physiol. Biochem.* **2010**, *48*, 636–645. [CrossRef]

54. Dong, W.; Xu, C.; Li, C.; Sun, J.; Zuo, Y.; Shi, S.; Cheng, T.; Guo, J.; Zhou, S. *ycf1*, the most promising plastid DNA barcode of land plants. *Sci. Rep.* **2015**, *5*, 8348. [CrossRef]

55. Strand, D.D.; D'Andrea, L.; Bock, R. The plastid NAD(P)H dehydrogenase-like complex: Structure, function and evolutionary dynamics. *Biochem. J.* **2019**, *476*, 2743–2756. [CrossRef]

56. Kemble, R.J. A rapid, single leaf, nucleic acid assay for determining the cytoplasmic organelle complement of rapeseed and related *Brassica* species. *Theor. Appl. Genet.* **1987**, *73*, 364–370. [CrossRef]

57. Scotti, N.; Cardi, T.; Marechal Drouard, L. Mitochondrial DNA and RNA isolation from small amounts of potato tissue. *Plant Mol. Biol. Rep.* **2001**, *19*, 67a–67h. [CrossRef]

58. Bolger, A.M.; Lohse, M.; Usadel, B. Trimmomatic: A flexible trimmer for Illumina sequence data. *Bioinformatics* **2014**, *30*, 2114–2120. [CrossRef]

59. Zerbino, D.R.; Birney, E. Velvet: Algorithms for *de novo* short read assembly using de Bruijn graphs. *Genome Res.* **2008**, *18*, 821–829. [CrossRef]

60. Assefa, T.; Mahama, A.A.; Brown, A.V.; Cannon, E.K.S.; Rubyogo, J.C.; Rao, I.M.; Blair, M.W.; Cannon, S.B. A review of breeding objectives, genomic resources, and marker-assisted methods in common bean (*Phaseolus vulgaris* L.). *Mol. Breed.* **2019**, *39*, 20. [CrossRef]

61. Langmead, B.; Salzberg, S.L. Fast gapped-read alignment with Bowtie 2. *Nat. Methods* **2012**, *9*, 357–359. [CrossRef]

62. Quinlan, A.R.; Hall, I.M. BEDTools: A flexible suite of utilities for comparing genomic features. *Bioinformatics* **2010**, *26*, 841–842. [CrossRef]

63. Weising, K.; Gardner, R.C. A set of conserved PCR primers for the analysis of simple sequence repeats polymorphisms in chloroplast genomes of dicotyledonous angiosperms. *Genome* **1999**, *42*, 9–19. [CrossRef]

64. Katoh, K.; Standley, D.M. MAFFT multiple sequence alignment software version 7: Improvements in performance and usability. *Mol. Biol. Evol.* **2013**, *30*, 772–780. [CrossRef]

65. Page, A.J.; Taylor, B.; Delaney, A.J.; Soares, J.; Seemann, T.; Keane, J.A.; Harris, S.R. Snp-sites: Rapid efficient extraction of SNPs from multi-fasta alignments. *Microb. Genom.* **2016**, *2*, e000056. [CrossRef]
66. Stamatakis, A. RAxML version 8: A tool for phylogenetic analysis and post-analysis of large phylogenies. *Bioinformatics* **2014**, *30*, 1312–1313. [CrossRef] [PubMed]

Publisher's Note: MDPI stays neutral with regard to jurisdictional claims in published maps and institutional affiliations.

Article

Transmission of Engineered Plastids in Sugarcane, a C$_4$ Monocotyledonous Plant, Reveals that Sorting of Preprogrammed Progenitor Cells Produce Heteroplasmy

Ghulam Mustafa and Muhammad Sarwar Khan *

Center of Agricultural Biochemistry and Biotechnology (CABB), University of Agriculture, University Road, Faisalabad, P.O. Box 38040, Pakistan; ghulam.mustafa@uaf.edu.pk
* Correspondence: sarwarkhan_40@hotmail.com; Tel.: +92-41-9201087

Abstract: We report here plastid transformation in sugarcane using biolistic transformation and embryogenesis-based regeneration approaches. Somatic embryos were developed from unfurled leaf sections, containing preprogrammed progenitor cells, to recover transformation events on antibiotic-containing regeneration medium. After developing a proficient regeneration system, the FLARE-S (fluorescent antibiotic resistance enzyme, spectinomycin and streptomycin) expression cassette that carries species-specific homologous sequence tails was used to transform plastids and track gene transmission and expression in sugarcane. Plants regenerated from streptomycin-resistant and genetically confirmed shoots were subjected to visual detection of the fluorescent enzyme using a fluorescent stereomicroscope, after genetic confirmation. The resultant heteroplasmic shoots remained to segregate on streptomycin-containing MS medium, referring to the unique pattern of division and sorting of cells in C$_4$ monocotyledonous compared to C$_3$ monocotyledonous and dicotyledonous plants since in sugarcane bundle sheath and mesophyll cells are distinct and sort independently after division. Hence, the transformation of either mesophyll or bundle sheath cells will develop heteroplasmic transgenic plants, suggesting the transformation of both types of cells. Whilst developed transgenic sugarcane plants are heteroplasmic, and selection-based regeneration protocol envisaging the role of division and sorting of cells in the purification of transplastomic demands further improvement, the study has established many parameters that may open up exciting possibilities to express genes of agricultural or pharmaceutical importance in sugarcane.

Keywords: plastid transformation; sugarcane; unfurled leaves; streptomycin; heteroplasmy; mesophyll and bundle sheath cells

Citation: Mustafa, G.; Khan, M.S. Transmission of Engineered Plastids in Sugarcane, a C$_4$ Monocotyledonous Plant, Reveals that Sorting of Preprogrammed Progenitor Cells Produce Heteroplasmy. *Plants* **2021**, *10*, 26. https://doi.org/10.3390/plants10010026

Received: 18 September 2020
Accepted: 14 November 2020
Published: 24 December 2020

Publisher's Note: MDPI stays neutral with regard to jurisdictional claims in published maps and institutional affiliations.

1. Introduction

Plastids have become attractive targets for genetic engineering since their genome offers several potential advantages, including accumulation of transporters to high levels with bona fide structures, biological containment of transgenes, gene stacking in operons, and absence of position effects [1,2]. Chloroplast genome has been engineered successfully to develop valuable traits, such as herbicide tolerance, insect and disease resistance, drought or salt tolerance, production of therapeutic proteins, antibodies, antibiotics, vaccine antigens, industrial enzymes, and other biosimilars in different plants [3–5]. Hence, more than 100 transgenes have been stably integrated and expressed in the chloroplast genome [6–8]. More recently, chloroplast genomes of major crops, including cotton, soybean, vegetables (carrot, lettuce, cabbage, eggplant, sugar beet), fruits (tomato), and trees (poplar and citrus), have been transformed [9,10].

Generally, monocots are recalcitrant to regeneration and there is no standard protocol available for efficient regeneration of plants from dedifferentiated cells, which could be used to transform plastids and to recover transgenic plastid-carrying cells. For each crop, different combinations of nutrients and auxins are used for callogenesis as well as

51

organogenesis. To date, plastid transformation has been reported in rice, yet the regenerated shoots remained heteroplasmic at plastome, organelle, and cell levels [11,12]. One of the major bottlenecks in developing stable plastid transformation in monocots has been their regeneration from non-green embryonic cells, containing undifferentiated plastids. Other barriers in developing homoplasmic transplastomic plants, particularly of rice, might be the low level of marker gene expression in non-green plastids in embryogenic cells because of a low genome copy number and low rates of protein synthesis [13].

While addressing bottlenecks, we attempted to transform the sugarcane plastid genome where we developed an efficient regeneration protocol and species-specific chloroplast transformation vector carrying the dominant visual selectable marker. The development of transplastomic technology in sugarcane will open up exciting possibilities for novel gene introduction and expression for agricultural or pharmaceutical traits.

2. Results

2.1. Development of a Proficient Regeneration Protocol

Unfurled leaves from six field-grown elite sugarcane genotypes (CPF-246, US-127, HSF-242, US-778, HSF-240, and US-64) were sacrificed to develop leaf roll discs. and subsequently placed on callus induction medium. Of six genotypes, maximum embryogenic cells were recovered from genotypes US-127, US-778 on MS medium supplemented with 2,4-D (2,4-Dichlorophenoxyacetic acid) ranging from 3 to 5 mg/L. Five-week-old dark proliferated calli were subjected to many combinations of hormones (RMOS, RMSD, RMSDB, and RMSDBK) (Section 4.2) and the maximum number of shoots was observed on RMSDBK medium (Figure 1). However, it was observed that maintenance of 2,4-D in combination with kinetin and BAP (6-Benzylaminopurine) results in healthy and separable mature plants, contrary to earlier studies (Figure 2) [14,15]. BAP, kinetin, and 2,4-D had synergistic effects on shoot regeneration [16,17].

Figure 1. Efficient multiple shoot regeneration from embryogenic calli developed from undifferentiated young leaf whorls of different cultivars of sugarcane on the RMSDBK medium. (**A**) US-778 (**B**) HSF-240 (**C**) US-64 (**D**) CPF-246 (**E**) US-127 (**F**) HSF-242.

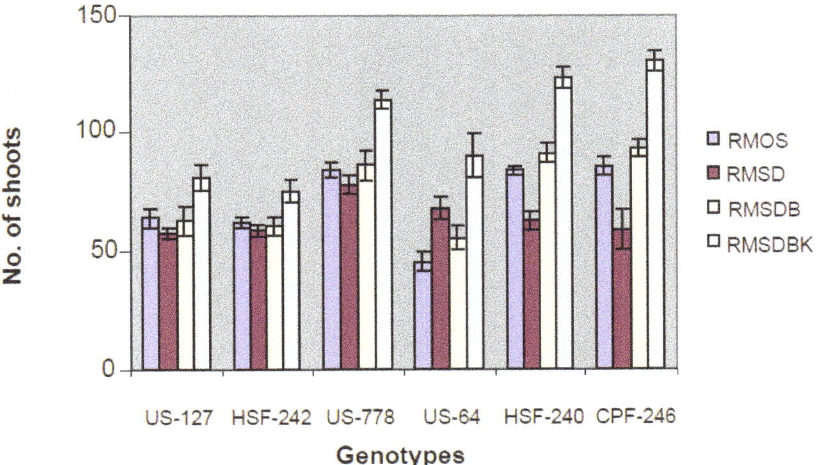

Figure 2. Effect of different types of media on shoot induction from calli derived from young leaf whorls of different genotypes. Most responsive genotype to regeneration was CPF-246 on RMSDBK medium.

2.2. Development of Species-Specific Chloroplast Transformation Vectors

After the establishment of a proficient regeneration system, a species-specific sugarcane plastid transformation vector was developed. Species-specific flanking sequences play a pivotal role in plastome engineering as a drastic decrease in transformation efficiency was observed when flanking sequences of petunia were used instead of tobacco for *Nicotiana* plastome transformation [18], suggesting that a lack of complete homology of the targeting sequences results in low transformation efficiency [19]. The stable plastid transformation system depends upon the integration of foreign DNA into the plastid genome through homologous recombination [20]. To facilitate homologous recombination, the inverted repeat region of sugarcane was amplified using primers P1 5′-GAT ATC AAA ACC CGT CCT CAG TTC GGA TTG C-3′ and P2 5′-GAT ATC CAC GAG TTG GAG ATA AGC GGA-3′. The *trnI*-*trn*A intergenic regions were selected to integrate transgene into the plastome since these regions have successfully been used to develop transgenic chloroplasts [21]. To engineer suitable restriction sites, for further cloning, an adapter sequence carrying restriction sites for *Bcu*I, *Kpn*I, *Apa*I, *Xho*I, *Hinc*II, *Bsu*15I, *Hind*III, *Eco*321, *Eco*RI, *Pst*I, *Sma*I, *Bam*HI, *Bcu*I, *Xba*I, *Not*I, *Bstx*I, and *Sac*I was cloned (Figure 3). The fluorescent selectable marker *FLARE-S* having *aadA* (aminoglycoside-3′-adenyltransferase) gene translationally fused with *gfp* (green fluorescent protein), was used to visually detect transformed cells since this is the most commonly used selection system giving out high transformation efficiency, developed by Khan and Maliga [11].

2.3. Plastid Transformation and Recovery of Transplastomic Plants

Selectable markers are essential tools for chloroplast transformation [11,22] and the only available dominant marker is *aadA*, conferring resistance to streptomycin and spectinomycin. Similar to other monocotyledonous cereals, sugarcane is naturally resistant to spectinomycin. Hence, streptomycin was used to select transformation events. After developing a kill curve, 300–350 mg/L streptomycin was determined as optimal for selection [23]. Three-week-old calli were bombarded using the PDS-1000/He biolistic gun (Bio-Rad, Hercules, CA, USA) following the established protocol [24]. The bombarded calli maintained on MS1.5 medium for 2–3 weeks were transferred onto streptomycin (350 mg/L)-containing RMSDBK medium where transformed cells proliferated and regenerated into shoots. However, wild-type calli turned brown and ultimately became dead

(Figure 4). GFP facilitated the selection and screening of the transformants. Calli chunks fluorescing green under UV light were separated from non-fluorescent cells and were regenerated into shoots. The regenerated shoots were shifted to the rooting medium. The antibiotic-resistant primary regenerants showed a great degree of phenotypic segregation during selection. Variegation was also observed in proliferating leaves along with green shoots. The green shoots appeared to be positive for marker gene presence and integration into plastome (Figure 5), inferring that the variegated shoots were highly heteroplasmic and transgenic plastids were segregating at the cell as well as tissue levels.

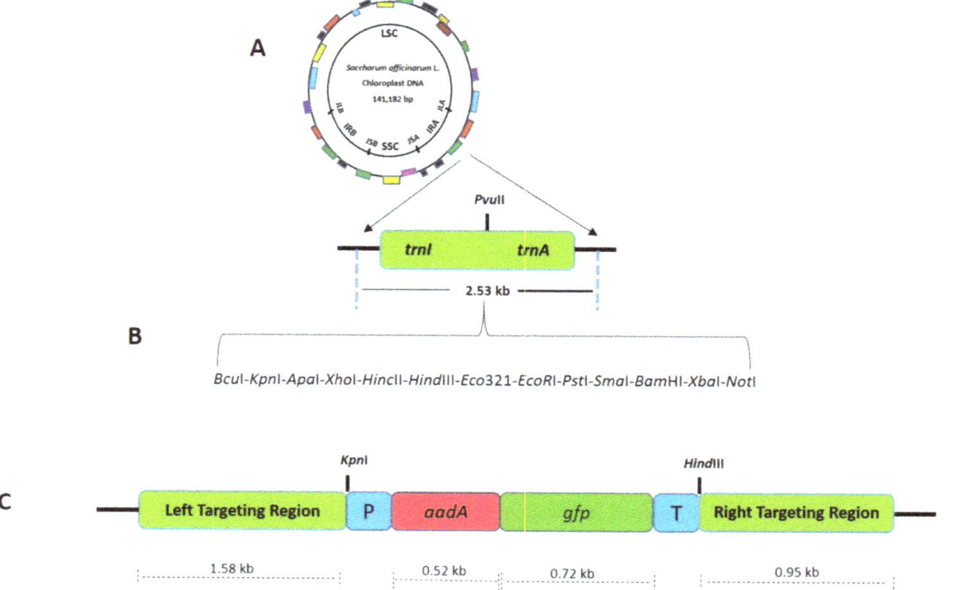

Figure 3. Schematic diagram showing the cloning strategy for the development of sugarcane plastid transformation vector SOFM2. (**A**) Sugarcane flanking sequences used for site-specific integration of transgene in the plastome inverted repeats. (**B**) Adapter sequence was cloned in between the flanking sequences to engineer multiple cloning sites. (**C**) Final sugarcane plastid transformation vector (SOFM2) with *aadA* and *gfp* genes.

2.4. Tracking Transgenic Plastids Using Green Fluorescence Protein (GFP)

The antibiotic-resistant calli and regenerated shoots were initially inspected for fluorescence using a hand-held long-wave UV lamp. The proliferating calli and leaves of the putative transplastomic sugarcane plants were subjected to a stereomicroscope, SZX (Olympus America, Melville, NY, USA) equipped with a GFP detection system. The green-fluorescing calli were selected and proliferated on streptomycin-containing MS medium supplemented with MS vitamins, 3 mg/L 2,4-D and were regenerated into shoots upon transfer to RMSDBK medium. Green-fluorescing sectors were observed in the leaves of antibiotic-resistant transgenic plants, confirming that the *gfp* gene was integrated into sugarcane plastome (Figure 6). These fluorescing sectors varied in size in leaves and even in different plants, depending upon the expression of the transgene as well as segregation of transplastomic and wild-type plastids (Figure 7). Such chimerism necessitates another cycle of regeneration on the selective medium. Accumulation of green fluorescent protein (GFP) was also assessed in the leaves of PCR-positive plants by confocal laser scanning microscopy. The FLARE-S expression confirmed the segregation of transplastomic and wild-type plastids in streptomycin-resistant sugarcane (Figure 8). It was observed that cells surrounding the veins are fluorescent, indicating chloroplasts in bundle sheath cells are only transformed, and regular files of these cell clones are extended in an aligned fashion.

Figure 4. Purification of shoots carrying transformed plastomes on streptomycin-containing regeneration medium. (**A**) Calli proliferated under dark growth conditions were used to transform plastomes (**B**) Selection of cells, after the bombardment, on streptomycin-containing regeneration medium where resistant cells are regenerated into shoots while sensitive are bleached. (**C**) Close up of the resistant shoots as shown at B. (**D**) Root initiation on streptomycin-containing rooting medium. (**E,F**) Multiplication of regenerated shoots on MS medium in jars and glass tubes with and without IBA, respectively.

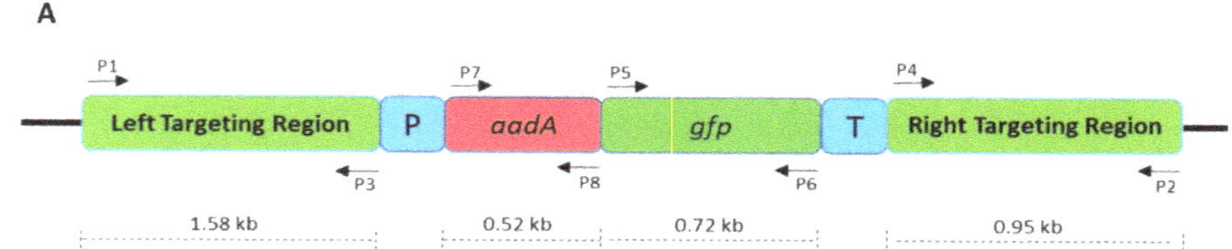

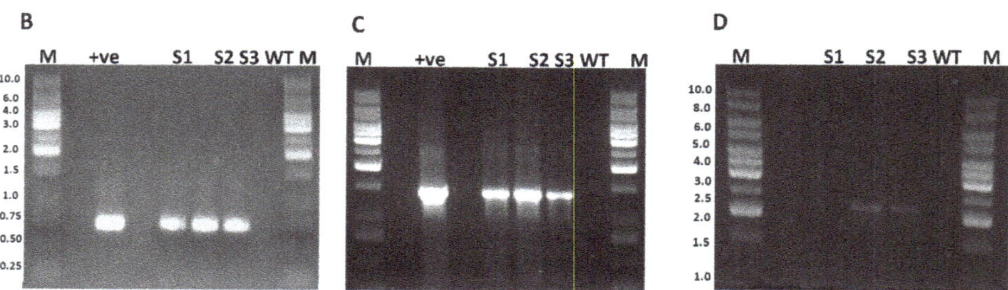

Figure 5. Confirmation of integration of FLARE-S cassette into the plastid genome of sugarcane. (**A**) Physical map of sugarcane plastid transformation vector showing the position of primers flanking various genes (**B**) PCR amplification of marker gene (*gfp*) with primer sets P5/P6: lanes M 1 kb DNA ladder, lane +ve shows amplification from plasmid DNA, lanes S1, S2, and S3 are transformed sugarcane plant DNA, lane WT untransformed sugarcane plant DNA (**C**) PCR amplification of FLARE-S with primer sets P6/P7: lanes M represent 1 kb DNA ladder, lane +ve represents plasmid DNA as template, lanes S1, S2, and S3 are transformed sugarcane plant DNA, lane WT untransformed sugarcane plant DNA (**D**) PCR amplification of left border sequence along with marker gene (*aadA*) with primer sets P1/P8: lanes S1, S2, and S3 are transformed sugarcane plant DNA, lane WT untransformed sugarcane plant DNA, lanes M 1 kb DNA ladder.

2.5. Tracking Transgene Integration through PCR Approach

The putative transgenic plants were analyzed for marker gene(s) integration into the plastid genome using the PCR approach. Two gene-specific primers for *gfp* were used. Clones 3, 5, and 6 were found positive for *gfp* gene integration since a fragment of 721 bp was amplified (Figure 5B). Further, to reconfirm the presence of translationally fused genes (*gfp* and *aadA*), another set of primers where forward primer landed on the *aadA* gene and reverse primer on the *gfp* gene were used. Amplification of a fragment of 1452 bp from clones 3, 5, and 6 confirmed the integration of the transgene into the plant genome. These experiments only verified the presence of transgenes in the plant genome but did not authenticate whether transgene was integrated into the plastome. To verify whether the transgenes are integrated into the chloroplast genome, another pair of primers was designed where one primer lands on native plastome and other on transgene (*aadA*). The amplification of a fragment of 2259 bp confirmed their transplasmicity, thus eliminating the possibility of spontaneous mutants, escapees, or even nuclear transformants.

However, plastid transformation was achieved in sugarcane, but low transformation efficiency and heteroplasmy remained to be resolved. The transformation efficiency resulting in one transformation event per 27 bombarded plates was found to be lower than reported in *Arabidopsis* [25,26], potato [27–29], and tomato [30,31]. However, transformation efficiency was comparable with rice [11] since target tissues in both cases were embryogenic calli. Plastids in the dark-grown calli are normally undeveloped, having a size of ~1 μm (approximately 5 to 10-fold smaller than well-developed chloroplasts in the green leaf tissue, turning out to be tough targets for transformation with 0.6 μm gold/tungsten particles owing to increased physical damage. The use of smaller-sized metal particles (0.4 μm) is reported to improve the transformation efficiency of proplastids by three- to four-fold [32]. Another reason could be the levels of transcription and translation that is lower in proplastids than mature chloroplasts [33]. The strong constitutive promoter (*Prrn*) was used to drive transgene expression but its activity has been reported to be low in proplastids. Thus, promoters encompassing both the PEP (plastid-encoded RNA polymerase) and NEP (nucleus-encoded RNA polymerase) would be the promoters of choice to increase transgene accumulation in sugarcane proplastids [34]. Another major impediment in achieving homoplasmic clones is the anatomy of the sugarcane plant where mesophyll and bundle sheath cells originate from preprogrammed dividing cells, and these cells are aligned in a regular fashion in the leaves.

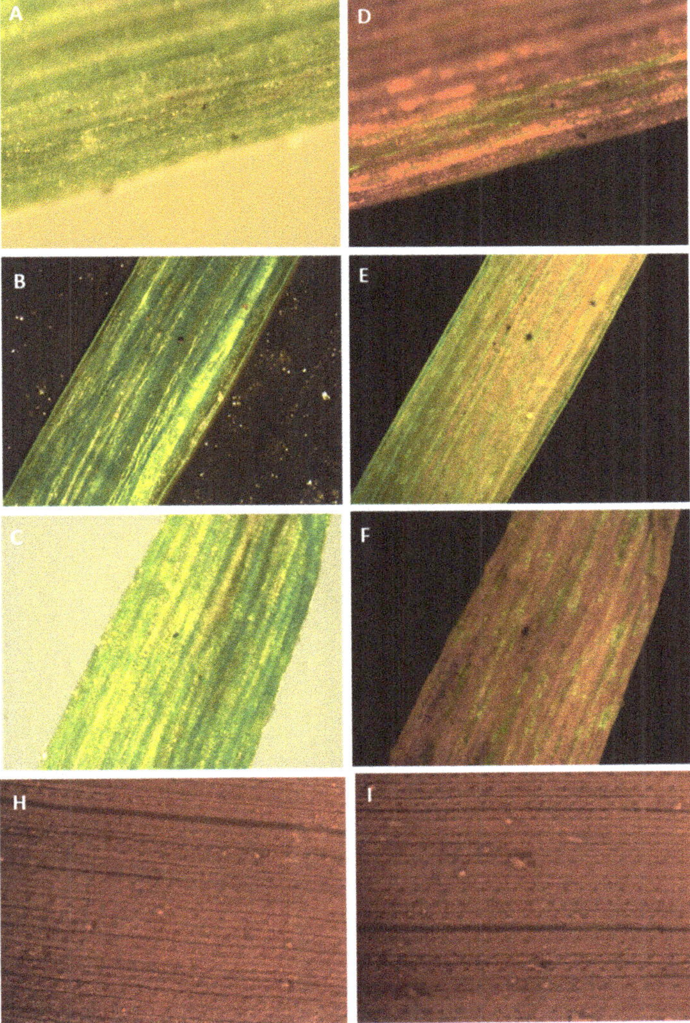

Figure 6. Tracking green fluorescent protein in sugarcane leaves by a stereomicroscope equipped with a GFP detection system. GFP fluoresces green whereas chlorophyll fluoresces red when exposed to fluorescent illumination. (**A–C**) Images of transformed sugarcane leaves in bright or dark field illumination. (**D–F**) Images of transformed sugarcane leaves in fluorescent illumination. (**H–I**) Images of untransformed sugarcane leaves in fluorescent illumination.

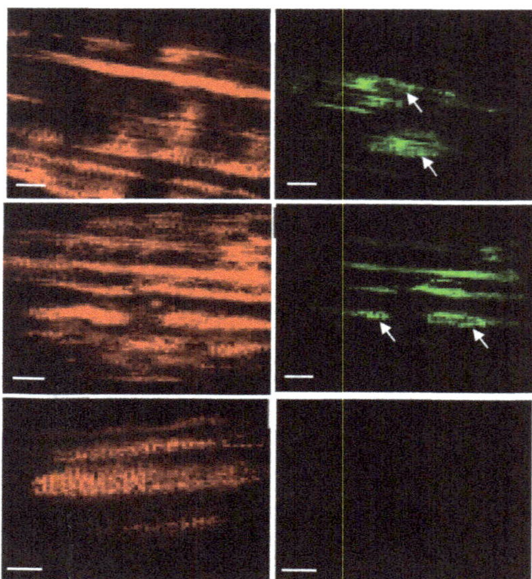

Figure 7. Visualizing fluorescent sectors in transformed sugarcane leaves under a laser scanning confocal microscope. The red channel and the green channel were imaged separately. Images were processed to detect chlorophyll (**A–C**) and Green Fluorescent Protein (**D–F**). Transplastomic plant leaf sectors fluoresced green (**D,E**) whereas no fluorescence was detected in untransformed sugarcane leaves (**F**) under the green channel.

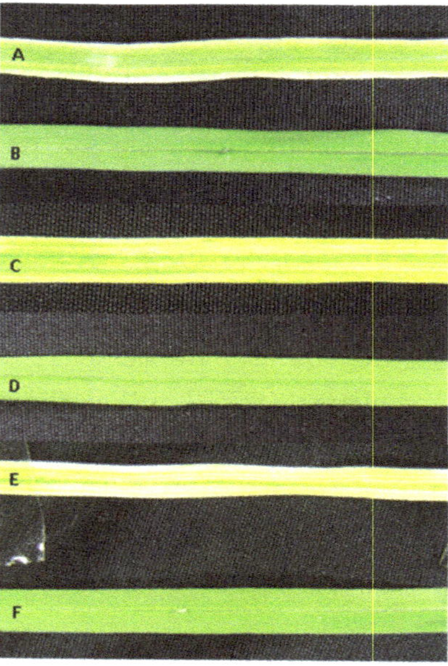

Figure 8. The phenotypic segregation of the antibiotic-resistant sugarcane transformants. Variegation was observed in leaves of plants growing on streptomycin-containing MS medium. (**A,C,E**) Variegated leaf segments of putative sugarcane transformants. (**B,D,F**) Lush green leaf segments of untransformed sugarcane plants growing on streptomycin-free medium in the same conditions.

3. Discussion

Plastid genome engineering is extended to many plant species, including tobacco, potato, cotton, tomato, carrot, oilseed rape, petunia, sugar beet, lettuce, cabbage, eggplant, and soybean, owing to the unique advantages of the expression of transgenes in polycistronic units, accumulation of transporters to high levels with bona fide structures, biological containment of transgenes because of maternal inheritance, and elimination of position effects that are frequently observed in nuclear transformation because of random insertion of transgenes into the genome. Plastid genome engineering is still very incipient in C_3 monocotyledonous crops, and it has not so far been reported for the C_4 sugarcane crop. Hence, the plastid genome of sugarcane is engineered. While designing experiments, three major bottlenecks to achieve homoplasmic transgenic plants were considered: one, referring to regeneration, the recalcitrance was thought to be the major bottleneck in plastome engineering of monocotyledonous crops as no mature leaf-based regeneration system is available. Therefore, many genotypes were used to assess the regeneration of cells/shoots on different combinations of hormones [35,36]. Consequently, a proficient regeneration system with higher regeneration efficiency than earlier reports [24,37] was developed.

The only dominant available marker for selection of plastid transformation events is the *aadA* gene that encodes aminoglycoside 3′-adenyltransferase and confers resistance to spectinomycin and streptomycin. Spectinomycin has been used extensively for selecting the transformation events in different plants like tobacco, potato, lettuce, tomato, cabbage, sugar beet, and eggplant, where green heteroplasmic shoots were recovered due to the phenotypic masking effect of transformed cells, subsequently purified through additional rounds of selection and sorting out of cells. Unlike these plants, rice [11] and sugarcane [23] are naturally resistant to spectinomycin. So, the second bottleneck was the choice of an antibiotic for selection and screening of transformation events on regeneration medium; and the only choice left was streptomycin. Hence, streptomycin was used in subsequent experiments after developing a kill curve [23]. Using streptomycin, transplastomic plants were recovered on the regeneration medium.

Third, the major impediment to plastid genome engineering in sugarcane observed was the presence of two preprogrammed distinct types of cells, bundle sheath and mesophyll, developed upon differentiation of meristematic cells. In C_4 plants, including maize, the bundle sheath (BS) lineage is distinct from that which produces mesophyll (M) cells [38]; this differentiation into the various cell types is driven by a complex interaction of transcription factors, small interfering RNAs, metabolites, and phytohormones [39]. Hence, the transformation of a proplastid in a non-differentiated or preprogrammed cell in the meristem tissue or calli may inherit transformed proplastid into a typical recipient cell type (either BS or M) and the transformed proplastids differentiate into chloroplasts leading towards heteroplasmy. This is because regular files of cell clones extend in an aligned fashion from the base to the tip of the leaf [40,41]. Emission of green fluorescence from aligned cells in leaves of regenerated streptomycin-resistant shoots (Figure 6) supports our hypothesis that a recipient cell of either BS or M upon transformation will divide and regular files of these cells will extend continuously in an aligned fashion in the leaves of a transgenic shoot. However, non-recipient cells will also grow similarly. Further, it is highly unlikely that transformed proplastids are inherited into both types of cells (mesophyll or bundle sheath), generating a homoplasmic plant. This, then, is one aspect that calls for further investigation.

4. Materials and Methods

4.1. Choice of Explant Material and Callus Induction

Various plant parts, including nodal tissues, apical meristem, the sub-apical region of the shoot apex, young leaf whorls, and root meristematic tissues, were used to develop calli and for proliferation. Release of phenolic compounds from all explants except young leaf whorls were a major problem in developing the calli and regeneration protocol, resulting in browning of tissues. Hence, the protocol for young leaf whorls was developed. Sugarcane tops of 6–8-month-old plants were sterilized using 70% ethanol and then sliced into transverse sections. Young leaf roll discs of 1.5–2.0 mm thickness and 60 to 95 mm^2 area cross-section were used to develop on callus induction medium, having different levels of 2,4-D (1, 2, 3, 4, and 5 mg/L). A suitable level of 2,4-D was selected that was tested in combination with different levels of kinetin (0.5, 1.0, 1.5, 2.0, 2.5, 3.0 mg/L).

4.2. Shoot Induction, Multiplication, and Rooting

Embryogenic calli were transferred to the RMOS medium. RMOS consists of Murashige and Skoog salts supplemented with thiamine HCl 1.0 mg/L, nicotinic acid 0.5 mg/L, pyridoxin HCl 0.5 mg/L, glycine 2 mg/L, myoinositol 100 mg/L, casein hydrolysate 500 mg/L, and sucrose 30 g/L. Afterward, calli were transferred to RMOS supplemented with 2,4-D (RMSD), 2,4-D and BAP (RMSDB) and 2,4-D, BAP, and kinetin (RMSDBK). The plates were incubated in low light (1400 lux day intensity) following 16:8 h light: dark regime for one week, and then in bright light (2000–2500 lux day intensity) at 26 ± 1 °C. The media were solidified using 2.6 g/L phytagel with a pH of 5.7. For multiplication and robust rooting, regenerated plants were transferred to the MSV medium (Murashige and Skoog's basal medium supplemented with IBA).

4.3. Construction of Species-Specific Chloroplast Transformation Vectors

Sugarcane species-specific plastid transformation vector SOFM2 was developed for targeted insertions into the *trn*I (97663–98685)-*trn*A (98751–99633) inverted repeat region of the plastome (GenBank accession No. NC 006084). The cloned fragment served as flanking borders in the final transformation vector. These flanking sequences (*trn*I-*trn*A) were amplified from the sugarcane plastid genome by PCR, using primers P1 (5′-GAT ATC AAA ACC CGT CCT CAG TTC GGA TTG C-3′) and P2 (5′-GAT ATC CAC GAG TTG GAG ATA AGC GGA-3′). P1 primer annealed to nucleotides 97101–97126 and P2 annealed to 99620–99641 of the plastome. A DNA fragment carrying restriction sites for endonucleases: *Bcu*I, *Kpn*I, *Apa*I, *Xho*I, *Hinc*II, *Bsu*15I, *Hind*III, *Eco*321, *Eco*RI, *Pst*I, *Sma*I, *Bam*HI, *Bcu*I, *Xba*I, *Not*I, *Bstx*I, and *Sac*I, was amplified by primers P3 (5′-GGT ACC ACT AGT GGG CCC CCC CTC GA-3′) and P4 (5′-GAG CTC CAC CGC GGT GGC GGC CGC T-3′) and cloned in the intergenic region to facilitate further cloning. The selection marker FLARE-S (fluorescent antibiotic resistance enzyme, spectinomycin, and streptomycin) encoding a bifunctional protein obtained by translational fusion of the aminoglycoside 3′-adenyltransferase (*aad*A) gene with the green fluorescent protein (*gfp*) gene was cloned at *Eco*RI/*Hind*III sites.

4.4. Sugarcane Plastome Transformation

The optimized conditions for efficient calli induction and regeneration were used to engineer the *Saccharum* plastome. Among the evaluated genotypes, genotype HSF-240 was selected for transformation owing to its good response to callogenesis and regeneration. Calli were bombarded with SOFM2 plasmid DNA-coated gold particles of 0.6 μm diameter, using a Biolistic gun PDS-1000/He (Bio-Rad, Hercules CA, USA), following the transformation procedures described by Khan and Maliga [11]. The calli were incubated in the dark at 26 ± 1 °C for 48 h before transfer to selective medium supplemented with 350 mg/L of streptomycin sulfate (Phytotechnology, Lenexa, KS, USA). After two weeks of incubation on callus induction medium, antibiotic-resistant calli were shifted to selective RMSDBK medium. The streptomycin-resistant shoots were rooted on MS medium containing 4 mg/L IBA.

4.5. Tracking Fluorescent Protein (gfp) in Plastids/Cells

Streptomycin-resistant calli were initially inspected for fluorescence using a hand-held long-wave UV lamp. The proliferating embryos and leaf segments of the putative transgenic sugarcane plants were subjected to fluorescence microscopy using a stereomicroscope equipped with GFP detection (Olympus SZX America, Melville, NY, USA). Sub-cellular localization of GFP was verified by a laser-scanning confocal microscope (Sarastro 2000

Confocal Image System; Molecular Dynamics, Sunnyvale, CA, USA). GFP fluorescence was detected in the FITC (Fluorescein isothiocyanate) channel (488–514 nm) whereas chlorophyll fluorescence was detected in the TRITC (Tetramethyl rhodamine isothiocyanate) channel (560–580 nm).

4.6. Total Cellular DNA Extraction and PCR Analyses

Total cellular DNA was isolated from leaf tissues of antibiotic-resistant sugarcane clones as well as from nontransformed plants. PCR analysis was carried out to validate transgene integration. *Taq* DNA polymerase (Fermentas, Hanover, MD, USA) and platinum *Taq* polymerase (Fermentas, Hanover, MD, USA) were used with 100–300 ng of genomic DNA as a template. The *gfp* gene was amplified with primers P5 (5′-CCA TGG CTA GTA AAG GAG AA-3′) and P6 (5′-TTA TTT GTA TAG TTC ATC CA-3′). The left border fragment was amplified with primers P1 (5′-GAT ATC AAA ACC CGT CCT CAG TTC GGA TTG C-3′) and P8 (5′-GGG CTG ATA CTG GGC CGG CAG G-3′).

5. Conclusions

A proficient regeneration protocol was developed for elite sugarcane genotypes. Using the protocol, species-specific plastid transformation vector, cell-autonomous selection, and visual detection approaches, plastid transformation was achieved in sugarcane. Streptomycin-resistant clones remained heteroplasmic after selection and regeneration on streptomycin-containing medium, referring to the complex pattern of division and sorting of cells in the meristematic tissues as another important factor in the purification of transplastomes to homoplasmy in sugarcane.

Author Contributions: M.S.K. designed experiments and managed funds for the said research. G.M. performed experiments under the supervision of M.S.K. All authors have read and agreed to the published version of the manuscript.

Funding: This research was funded by the Ministry of Science & Technology (MoST), Islamabad, Pakistan to M.S.K.

Institutional Review Board Statement: Not applicable.

Informed Consent Statement: Not applicable.

Data Availability Statement: Not applicable.

Acknowledgments: The authors would like to thank the Ministry of Science & Technology, Pakistan for providing funds; and Sugarcane Research Institute, AARI Faisalabad, Pakistan for providing the plant material.

Conflicts of Interest: The authors declare no conflict of interest relevant to the publication of this manuscript.

References

1. Khan, M.S. Plastid genome engineering in plants: Present status and future Trends. *Mol. Plant Breed.* **2012**, *8*, 91–102. [CrossRef]
2. Khan, M S.; Kanwal, B.; Nazir, S. Metabolic engineering of the chloroplast genome reveals that the yeast *ArDH* gene confers enhanced tolerance to salinity and drought in plants. *Front. Plant Sci.* **2015**, *6*, 725. [CrossRef] [PubMed]
3. Clarke, J.L.; Daniell, H. Plastid biotechnology for crop production: Present status and future perspectives. *Plant Mol. Biol.* **2011**, *76*, 207–209. [CrossRef] [PubMed]
4. Ancin, M.; Sanz-Barrio, R.; Santamaria, E.; Millan, A.F.; Larraya, L.; Veramendi, J.; Farran, I. Functional improvement of human cardiotrophin 1 produced in tobacco chloroplasts by co-expression with plastid thioredoxin m. *Plants* **2020**, *9*, 183. [CrossRef] [PubMed]
5. Fuentes, P.; Armarego-Marriott, T.; Bock, R. Plastid transformation and its application in metabolic engineering. *Curr. Opin. Biotechnol.* **2018**, *49*, 10–15. [CrossRef] [PubMed]
6. Bock, R.; Warzecha, H. Solar-powered factories for new vaccines and antibiotics. *Trends Biotechnol.* **2010**, *28*, 246–252. [CrossRef] [PubMed]
7. Bock, R. Engineering plastid genomes: Methods, tools, and applications in basic research and biotechnology. *Annu. Rev. Plant Biol.* **2015**, *66*, 211–241. [CrossRef]

8. Gan, Q.; Jiang, J.; Han, X.; Wang, S.; Lu, Y. Engineering the chloroplast genome of oleaginous marine microalga *Nannochloropsis oceanica*. *Front. Plant Sci.* **2018**, *9*, 439. [CrossRef]

9. Verma, D.; Daniell, H. Chloroplast vector systems for biotechnology applications. *Plant Physiol.* **2007**, *145*, 1129–1143. [CrossRef]

10. Khan, M.S.; Mustafa, G.; Joyia, F.A. Technical Advances in Chloroplast Biotechnology. In *Transgenic Crops*; Khan, M.S., Ed.; InTech-Open Science: London, UK, 2019; pp. 1–13. [CrossRef]

11. Khan, M.S.; Maliga, P. Fluorescent antibiotic resistance marker for tracking plastid transformation in higher plants. *Nat. Biotechnol.* **1999**, *17*, 910–915. [CrossRef]

12. Lee, S.M.; Kang, K.; Chung, H.; Yoo, S.H.; Xu, X.M.; Lee, S.B.; Cheong, J.J.; Daniell, H.; Kim, M. Plastid transformation in the monocotyledonous cereal crop, rice (*Oryza sativa*) and transmission of transgenes to their progeny. *Mol. Cells* **2006**, *21*, 401–410. [PubMed]

13. Daniell, H.; Khan, M.S.; Allison, L. Milestones in chloroplast genetic engineering: An environmentally friendly era in biotechnology. *Trends Plant Sci.* **2002**, *7*, 84–91. [CrossRef]

14. Biradar, S.; Biradar, D.P.; Patil, V.C.; Patil, S.S.; Kambar, N.S. In vitro plant regeneration using shoot tip culture in commercial cultivar of sugarcane. *Karnat. J. Agri. Sci.* **2009**, *22*, 21–24.

15. Chen, W.H.; Davey, M.R.; Power, J.B.; Cocking, E.C. Control and maintenance of plant regeneration in sugarcane callus cultures. *J. Exp. Bot.* **1988**, *39*, 251–261. [CrossRef]

16. Ather, A.; Khan, S.; Rehman, A.; Nazir, M. Optimization of the protocols for callus induction, regeneration and acclimatization of sugarcane cv. Thatta-10. *Pak. J. Bot.* **2009**, *41*, 815–820.

17. Mustafa, G.; Khan, M.S. Reproducible in vitro regeneration system for purifying sugarcane clones. *Afr. J. Biotechnol.* **2012**, *11*, 9961–9969. [CrossRef]

18. Fernandez-San, M.A.; Mingo-Castel, A.; Miller, M.; Daniell, H. A chloroplast transgenic approach to hyper express and purify human serum albumin, a protein highly susceptible to proteolytic degradation. *Plant Biotechnol.* **2003**, *1*, 77–79. [CrossRef]

19. Kavanagh, T.A.; Thanh, N.D.; Lao, N.T.; Grath, N.M.; Peter, S.O.; Horvath, E.M.; Dix, P.J.; Medgyesy, P. Homologous plastid DNA transformation in tobacco is mediated by multiple recombination events. *Genetics* **1999**, *152*, 1111–1122. [PubMed]

20. Khan, M.S.; Khalid, A.M.; Malik, K.A. Intein-mediated protein trans-splicing and transgene containment in plastids. *Trends Biotechnol.* **2005**, *23*, 217–220. [CrossRef]

21. Dhingra, A.; Portis, A., Jr.; Daniell, H. Enhanced translation of a chloroplast-expressed *RbcS* gene restores small subunit levels and photosynthesis in nuclear *RbcS* antisense plants. *Proc. Natl. Acad. Sci. USA* **2004**, *101*, 6315–6320. [CrossRef]

22. Day, A.; Goldschmidt-Clermont, M. The chloroplast transformation toolbox: Selectable markers and marker removal. *Plant Biotechnol. J.* **2011**, *9*, 540–553. [CrossRef] [PubMed]

23. Mustafa, G.; Khan, M.S. Prospecting the utility of antibiotics as lethal selection agents for chloroplast transformation of sugarcane. *Int. J. Agri. Biol.* **2012**, *14*, 307–310.

24. Khan, M.S.; Ali, S.; Iqbal, J. Developmental and photosynthetic regulation of δ-endotoxin reveals that engineered sugarcane conferring resistance to 'dead heart' contains no toxins in cane juice. *Mol. Biol. Rep.* **2011**, *38*, 2359–2369. [CrossRef] [PubMed]

25. Sikdar, S.R.; Serino, G.; Chaudhuri, S.; Maliga, P. Plastid transformation in *Arabidopsis thaliana*. *Plant Cell Rep.* **1998**, *18*, 20–24. [CrossRef]

26. Yu, Q.; Lutz, K.A.; Maliga, P. Efficient plastid transformation in Arabidopsis. *Plant Physiol.* **2017**, *175*, 186–193. [CrossRef]

27. Sidorov, V.A.; Kasten, D.; Pang, S.Z.; Hajdukiewicz, P.T.J.; Staub, J.M.; Nehra, N.S. Stable chloroplast transformation in potato: Use of green fluorescent protein as a plastid marker. *Plant J.* **1999**, *19*, 209–216. [CrossRef]

28. Valkov, V.T.; Gargano, D.; Manna, C.; Formisano, G.; Dix, P.J.; Gray, J.C.; Scotti, N.; Cardi, T. High efficiency plastid transformation in potato and regulation of transgene expression in leaves and tubers by alternative 5′ and 3′ regulatory sequences. *Transgenic Res.* **2011**, *20*, 137–151. [CrossRef]

29. Valkov, V.T.; Gargano, D.; Scotti, N.; Cardi, T. Plastid transformation in potato: *Solanum tuberosum*. *Methods Mol. Biol.* **2014**, *1132*, 295–303. [CrossRef]

30. Ruf, S.; Hermann, M.; Berger, I.J.; Carrer, H.; Bock, R. Stable genetic transformation of tomato plastids and expression of a foreign protein in fruit. *Nat. Biotechnol.* **2001**, *19*, 870–875. [CrossRef]

31. Ruf, S.; Bock, R. Plastid transformation in tomato. *Methods Mol. Biol.* **2014**, *1132*, 265–276. [CrossRef]

32. Langbecker, C.L.; Ye, G.N.; Broyles, D.L.; Duggan, L.L.; Xu, C.W.; Hajdukiewicz, P.T.J.; Armstrong, C.L.; Staub, J.M. High-frequency transformation of undeveloped plastids in tobacco suspension cells. *Plant Physiol.* **2004**, *135*, 39–46. [CrossRef] [PubMed]

33. Silhavy, D.; Maliga, P. Mapping of promoters for the nucleus-encoded plastid RNA polymerase (NEP) in the iojap maize mutant. *Curr. Genet.* **1998**, *33*, 340–344. [CrossRef] [PubMed]

34. Maliga, P. Progress towards commercialization of plastid transformation technology. *Trends Biotechnol.* **2003**, *21*, 20–28. [CrossRef]

35. Gill, N.K.; Gill, R.; Gosal, S.S. Factors enhancing somatic embryogenesis and plant regeneration in sugarcane (Saccharum officinarum L.). *Ind. J. Biotechnol.* **2004**, *3*, 119–123.

36. Gallo-Meagher, M.; English, R.G.; Abouzid, A. Thidiazuron stimulates shoot regeneration of sugarcane embryogenic callus. *Vitro Cell. Dev. Biol. Plant* **2000**, *36*, 37–40. [CrossRef]

37. Ali, S.; Iqbal, J.; Khan, M.S. Genotype independent in vitro regeneration system in elite varieties of sugarcane. *Pak. J. Bot.* **2010**, *42*, 3783–3790.

38. Sud, R.M.; Dengler, N.G. Cell lineage of vein formation in variegated leaves of the C$_4$ grass *Stenotaphrum secundatum*. *Ann. Bot.* **2000**, *86*, 99–112. [CrossRef]
39. Byrne, M.E. Making leaves. *Curr. Opin. Plant Biol.* **2012**, *15*, 24–30. [CrossRef]
40. Laetsch, W.M.; Price, I. Development of the dimorphic chloroplasts of sugarcane source. *Am. J. Bot.* **1969**, *56*, 77–87. [CrossRef]
41. Poethig, R.S. Cellular parameters of leaf morphogenesis in maize and tobacco. In *Contemporary Problems in Plant Anatomy*; White, R.A., Dickison, W.C., Eds.; Academic Press: New York, NY, USA, 1984; pp. 235–259.

Article

A Core Module of Nuclear Genes Regulated by Biogenic Retrograde Signals from Plastids

Björn Grübler, Carolina Cozzi and Thomas Pfannschmidt *

Pflanzenphysiologie, Institut für Botanik, Naturwissenschaftliche Fakultät, Leibniz-Universität Hannover, Herrenhäuser Str. 2, 30419 Hannover, Germany; BjoernGruebler@gmx.de (B.G.); c.cozzi@botanik.uni-hannover.de (C.C.)
* Correspondence: t.pfannschmidt@botanik.uni-hannover.de

Abstract: Chloroplast biogenesis during seedling development of angiosperms is a rapid and highly dynamic process that parallels the light-dependent photomorphogenic programme. Pre-treatments of dark-grown seedlings with lincomyin or norflurazon prevent chloroplast biogenesis upon illumination yielding albino seedlings. A comparable phenotype was found for the *Arabidopsis* mutant *plastid-encoded polymerase associated protein 7* (*pap7*) being defective in the prokaryotic-type plastid RNA polymerase. In all three cases the defect in plastid function has a severe impact on the expression of nuclear genes representing the influence of retrograde signaling pathway(s) from the plastid. We performed a meta-analysis of recently published genome-wide expression studies that investigated the impact of the aforementioned chemical and genetic blocking of chloroplast biogenesis on nuclear gene expression profiles. We identified a core module of 152 genes being affected in all three conditions. These genes were classified according to their function and analyzed with respect to their implication in retrograde signaling and chloroplast biogenesis. Our study uncovers novel genes regulated by retrograde biogenic signals and suggests the action of a common signaling pathway that is used by signals originating from plastid transcription, translation and oxidative stress.

Keywords: plastids; photomorphogenesis; retrograde control; biogenic signals; lincomycin; norflurazon; *pap7-1* mutant

Citation: Grübler, B.; Cozzi, C.; Pfannschmidt, T. A Core Module of Nuclear Genes Regulated by Biogenic Retrograde Signals from Plastids. *Plants* **2021**, *10*, 296. https://doi.org/10.3390/plants10020296

Academic Editor: Nunzia Scotti
Received: 18 January 2021
Accepted: 29 January 2021
Published: 4 February 2021

Publisher's Note: MDPI stays neutral with regard to jurisdictional claims in published maps and institutional affiliations.

1. Introduction

Chloroplasts are sub-cellular organelles in plants and algae that perform photosynthesis and many other metabolic activities. In angiosperms they develop from undifferentiated precursors called proplastids which are inherited from the mother plant in the cells of the embryo [1,2]. Upon germination, the embryo develops into a rapidly growing seedling. In light this development follows a photomorphogenic programme which includes opening of the cotyledons, repression of hypocotyl elongation and greening. The latter is due to the biogenesis of chloroplasts from the proplastids in the cotyledons. However, in the case that germination occurs in the dark because the seed is buried by humus or soil, the seedling follows a different developmental programme called skotomorphogenesis [3]. Here, the cotyledons remain small and of yellow color without any expansion. They are directed downwards by an apical hook which protects the apical meristem while the highly elongating hypocotyl drives the cotyledons towards the soil surface. The proplastids in this etiolated seedling develop into yellow etioplasts, an intermediate developmental stage of plastids incapable of performing photosynthesis. However, etioplasts develop into chloroplasts within hours as soon as the germinating seedling perceives light. Etioplast-to-chloroplast conversion, thus, is often used as experimental system to study molecular basics of chloroplast development [4–7].

The molecular steps controlling the biogenesis of chloroplasts are far from understood, mostly because of the rapidity and complexity of the processes involved [7]. Initiation of photomorphogenesis, starting either directly from proplastids or from etioplasts, occurs by the activation of the phytochrome system through light. It largely determines the morphological changes of the seedling when it enters photomorphogenesis. Chloroplast biogenesis occurs at the same time; however, many observations in recent years indicate that this developmental process is rather a parallel than an intrinsic part of photomorphogenesis [8]. For instance *constitutive photomorphogenesis (cop)* mutants from *Arabidopsis* develop a photomorphogenic phenotype in the dark without chloroplast formation [9]. Vice versa, *plastid-encoded RNA polymerase-associated protein (pap)* mutants develop a normal photomorphogenic phenotype in the light without chloroplast formation [10]. Chloroplast biogenesis, therefore, is neither a prerequisite nor a consequence of photomorphogenesis and it remains to be elucidated how it is connected to the photomorphogenic programme.

The major steps of chloroplast biogenesis involve the build-up of the internal thylakoid membrane system and the assembly of the photosynthetic apparatus. Because of their endosymbiotic ancestry chloroplasts possess their own genome that encodes central components of the photosynthetic and gene expression machineries [11]. However, assembly of functional membrane structures and protein complexes requires the import of thousands of nuclear-encoded components. Plastids, thus, are regarded as genetically semi-autonomous. Because of the high copy-number of the plastid genome and the fact, that each cell contains many plastids, one can observe a strong imbalance in the ratio of plastid over nuclear genes encoding plastid proteins. Proper timely and spatial expression of genes essential for chloroplast biogenesis, therefore, requires a high coordination between the two genetic compartments. This is achieved by a mutual information exchange called anterograde signaling (nucleus-to-plastid signaling) and retrograde signaling (plastid-to nucleus signaling) [12–17].

Retrograde signals from plastids during early steps of chloroplast biogenesis have been named biogenic signals (in contrast to operational signals from fully active chloroplasts) [18]. These signals were discovered in experiments where plastid development was either chemically or genetically inhibited resulting in a parallel inhibition of the expression of nuclear genes encoding plastid photosynthesis proteins such as subunits of the ribulose-bis-phosphate carboxylase/oxygenase (Rubisco) or light harvesting complexes (Lhc) of the photosystems [19,20]. This led to the concept of a plastid factor or signal that is required for the proper development of the chloroplast [21]. This research field has seen a tremendous effort in the last decade and it became clear that biogenic signals from plastids likely play an important role in the regulation of chloroplast biogenesis. How they are implicated in detail, is however, far from understood [22].

Two common approaches for the study of biogenic signals are the treatments of germinating seedlings with norflurazon (NF) and lincomycin (LIN). NF is an inhibitor of the plastid phytoene desaturase and a potent repressor of carotenoid biosynthesis. NF-treated dark-grown seedlings remain completely white and experience a severe oxidative stress from photo-sensitization of protochlorophyllide upon illumination since the quenching properties of the carotenoids are missing. The generated reactive oxygen species block any further steps towards chloroplast biogenesis by oxidative destruction of the internal plastid structures [23]. LIN treatment, in contrast, blocks the plastid translation machinery and prevents the build-up of all plastid-encoded protein components of the photosynthesis apparatus including the core proteins of the photosystems which are essential for a proper assembly of the systems [24]. Both treatments act on different sites in the plastid, but the resulting phenotypes both at phenotypic and molecular levels are similar in many aspects including an albino appearance and a repression of nuclear encoded photosynthesis associated nuclear genes (PhANGs).

Recently we reported a detailed expression profiling of the *pap7-1* mutant of *Arabidopsis* [25]. This mutant displays a defect in the activity of the plastid encoded RNA polymerase (PEP) and exhibits a severe disturbance in plastid and nuclear gene expression leading to an albino phenotype. Surprisingly, the repressive impact on PhANG expression in this mutant was rather weak and strong repression was found to be limited to the group of *Lhc* genes contrasting the notion that inhibition of plastid development causes a general repression of PhANGs. Therefore, we were wondering what the "true" impact of biogenic signals on chloroplast biogenesis is. Here, we present a meta-analysis of gene expression profiles obtained from the *Arabidopsis pap7-1* mutant and two recent gene expression studies using NF and LIN in order to define commonalities and differences between the three approaches. Our data uncover a core gene module that exhibits common expression profiles in all three conditions and identify potential novel targets of biogenic signals as well as regulators of chloroplast biogenesis.

2. Material and Methods

2.1. Sets of Retrograde Controlled Genes Used in This Study

Primary expression data sets for genes being controlled by retrograde signals identified under influence of NF and LIN treatments were taken from previously published studies [26,27]. Data sets for retrograde controlled genes in the *pap7-1* mutant as well as for light-controlled genes in wild-type plants were taken from our own data sets [25]. The study on LIN effects was performed as a kinetic experiment with samples taken 0.5, 1, 4 and 24 h after a light intensity shift. In pre-selection comparisons we observed that the number of overlapping genes between the *pap7-1* and LIN data sets increased with time. The 24 h LIN data set, therefore, was used as base for further meta-analysis. Only genes that exhibited a significant relative expression change (repression or induction) in response to plastid dysfunction of at least log2 $\geq$ 1 and a *p*-value $\leq$ 0.05 in all data sets were included in our comparisons.

2.2. Comparison of Gene Lists

For all principal comparisons of microarray data sets and the identification of overlapping genes between the three studies we used standard functions of Microsoft EXCEL. The corresponding Venn diagram was generated by using a web-based tool for comparison of large data sets (InteractiVenn; http://www.interactivenn.net/index.html) [28].

2.3. Functional Annotation and Localization

The initial functional annotation of the identified 152 genes is based on the MapMan Bin categories [29] and was curated manually for gene descriptions and functions of encoded proteins afterwards. To this end each gene was checked for database entries in The Arabidopsis Information Resource (TAIR) (https://www.arabidopsis.org) and The Universal Protein Resourcce (UniProt) (https://www.uniprot.org). Information on potential interaction partners was obtained from the Biological General Repository for Interaction Datasets (BioGRID) (https://thebiogrid.org) and the real-time multiple association network integration algorithm for predicting gene function (GeneMANIA) (https://genemania.org) databases. In addition, information on potential localization of gene products was extracted from the Bio-Analytic Resource (BAR) for Plant Biology and the integrated Arabidopsis Cell eFP Browser (https://bar.utoronto.ca) and aligned with the corresponding information in the Arabidopsis thaliana chloroplast protein database AT_CHLORO (http://at-chloro.prabi.fr) for sub-plastidial localization [30]. Genes were then classified according to major functional categories (which may differ in some cases from the original MapMan bins) and were given as heat map sorted on the base of the expression values in the LIN data set. All collected information about the gene products are summarized in Supplemental Table S1.

3. Results and Discussion

3.1. Selection and Comparability of Microarray Data Sets

Recently, we identified a set of retrograde controlled genes that are repressed or induced by biogenic signals in the light when chloroplast biogenesis is blocked at the level of plastid transcription [25]. We were interested in understanding how similar or different these gene groups are in comparison to conditions when the block of chloroplast biogenesis occurs at the level of plastid translation or at a level of general destruction through oxidative damage. To this end we compared our data with results from two investigating the effects of LIN and NF on nuclear gene expression [26,27]. The experimental design of all three studies was highly similar using *Arabidopsis* seedlings grown for 5–7 days on sugar-supplemented medium in Petri dishes and exposed to light to visualize the effect of blocked chloroplast biogenesis on light-regulated gene expression profiles. In LIN-based experiments seedlings were grown in absence and presence of 0.5 mM LIN under extremely weak blue-red light of 0.5 µE fluence rate for 6 days followed by a shift to 60 µE blue-red light for 24 h [27]. In the NF experiment seedlings were grown in absence and presence of 5 µM NF for 3 days in the dark followed by 3 days in white light [26]. In both studies the inhibitors, thus, had sufficient time to block the respective process before induction of chloroplast biogenesis. In our own experiments with the genetically blocked *pap7-1* mutant seedlings were grown for 5 days directly under light since the genetic inactivation had become effective already during the establishment of the seeds [25]. Therefore, in all three experiments the effect of a block in chloroplast biogenesis on light-controlled gene expression was studied at a comparable developmental stage of the seedlings (2 cotyledons stage). Furthermore, all studies used the Affymetrix ATH1 *Arabidopsis* chips providing high technical comparability. We, therefore, regarded the experimental set-ups as sufficiently similar to provide reliable data for a meta-analysis.

3.2. Identification of a Core-Module of Genes Controlled by Biogenic Signals

The chosen gene sets contained 3004 genes for the LIN set, 1957 genes for the *pap7-1* set and 1128 genes for the NF set. Comparison of the three conditions revealed a common gene module of 152 genes being significantly regulated in all three conditions (Figure 1A). The highest bivalent similarity was found for the NF set that shared 51.1% of its genes with the LIN set and 21.5% with the *pap7-1* set. The *pap7-1* set shared 25.8% of its genes with the NF set and 12.4% with the LIN set while the LIN set shared 19.2% of its genes with the NF set and 16.8% with the *pap7-1* set (Figure 1B). These results strongly suggest that each of the three different blocks in chloroplast biogenesis induce largely their own specific responses in nuclear gene expression. These differences in expression are likely attributed to the different sites of the respective blocks and/or the slightly different experimental set-ups and laboratory conditions. Nonetheless, there apparently exists also a common and robust element in all three conditions giving rise to the shared group of 152 genes. Therefore, we regard this group as a highly trustful core module of genes being influenced in their expression when chloroplast biogenesis is blocked, thus representing genuine target genes regulated (directly or indirectly) by retrograde biogenic (RB) signals.

An initial functional annotation of the 152 genes based on the bin definition in MapMan [31] was performed in order to understand the regulatory implications of the biogenic control. The largest group comprised 38 genes encoding proteins of unassigned, unknown or hypothetical function (Figure 1C) strongly suggesting that many aspects of chloroplast biogenesis at the molecular level are not understood yet. The second largest group contained 18 genes encoding components implicated in "Photosynthesis" followed by groups involved in Protein" (15), "RNA" (11), "Redox" (10), "Transport" (9), "Hormone metabolism" (6), "Stress" (5) and 14 further groups represented by one to four genes.

The expression behavior of the 152 genes was highly similar in the three analyzed conditions with approximately one third (47) of genes being stronger expressed than in unaffected wild-type (WT) controls and two third (105) of genes exhibiting a weaker expression (Supplemental Table S1). In control conditions detecting the influence of light (WT-light versus WT-dark) most of the genes displaying a weaker expression upon the block in chloroplast biogenesis exhibited an opposite response with a clear induction upon illumination (and vice versa). This opposite expression pattern corresponds to recent reports where this pattern was interpreted in a way that biogenic signals turn light signals from positive into negative stimuli [32]. We, therefore, compared the core module to a set of recently identified direct HY5 target genes [33]. HY5 is one of the central regulators in photomorphogenesis and expected to control or co-regulate a large part of the light regulated genes [34] during seedling photomorphogenesis and chloroplast biogenesis. However, we identified only 5 of the 152 genes in the set of HY5-dependent genes (Supplemental Figure S1) indicating that the core module is likely not directly regulated by the classical light-dependent signaling pathways.

Except for a few condition-specific expressed genes (see below), the general profiles of genes affected by LIN and NF treatments were found to be highly similar in strength and direction of expression change. The expression profiles in the *pap7-1* mutant were comparable to both LIN and NF profile concerning the direction of expression change; however, the degree of variation was weaker in most cases. Pharmaceutical inhibition of chloroplast development by LIN or NF treatments, thus, appears to have a stronger impact on nuclear gene expression than the genetic block in the *pap7-1* mutant, although the overall plant phenotype in all three conditions was largely the same. This suggests that the inhibitor treatments (i) induce stronger inhibitory effects in the plastid than the genetic block and/or (ii) the respectively affected plastid processes contribute to biogenic retrograde signaling with different strength. This suggest that the albino phenotype *per se* is likely not the cause of the changes in nuclear gene expression arguing for a specific molecular signaling pathway.

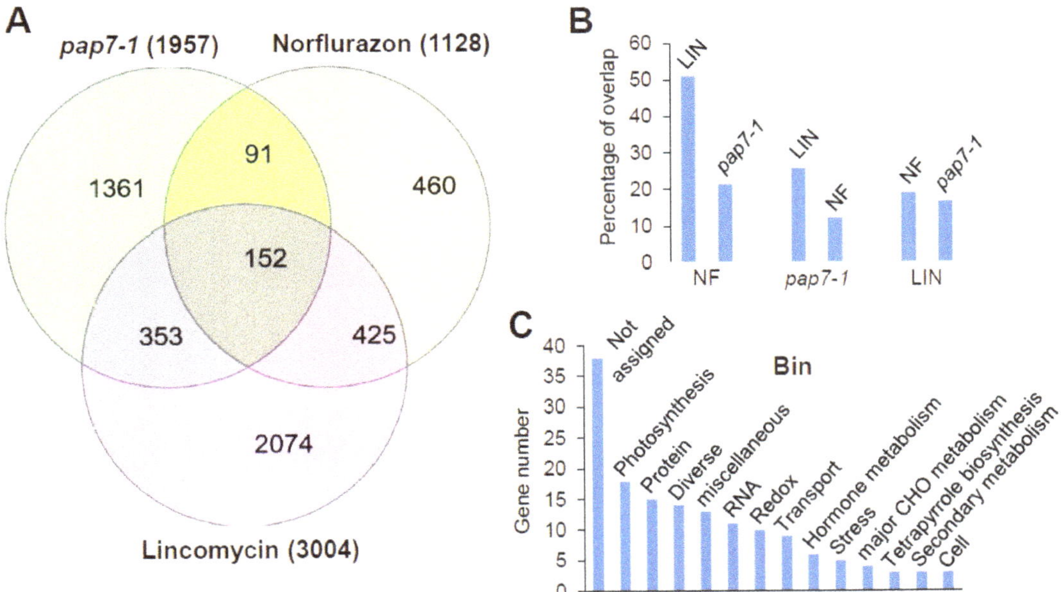

Figure 1. Comparison of target gene modules identified in three experimentally different approaches blocking chloroplast biogenesis. (**A**) Venn diagram describing the overlap in target genes of biogenic signals identified by norflurazon (NF) treatment, 24 h lincomycin (LIN) treatment and by genetic inactivation in *pap7-1* mutants. In all studies significantly regulated genes were defined by an expression change of at least log2 fold change ≥ 1 and a p-value ≤ 0.05. (**B**) Percentage of shared genes between the treatments. Since the sizes of the affected gene groups in the three conditions are different, the overlap is given for each of the three binoms (indicated on bottom) separately. (**C**) Functional annotation of genes shared between all three conditions. The categorization followed the Bin system of MapMan. Gene groups with a minimum of at least three genes are mentioned. A complete list of all genes is given in Supplemental Table S1.

3.3. Functional Subsets within the Core Module

Because of the high number of genes encoding proteins with unknown or unassigned functions, we curated the functional annotation of each gene in the identified core module manually using current databases and literature and generated an up-dated list that covers detailed information on function, potential interaction partners and intracellular localization (Supplemental Table S1). By this means we uncovered a number of interesting novel targets for RB signals not yet described (see description of functional subsets below).

In total 84 of the 152 genes encode proteins that were predicted to be localized with low, medium or high probability in plastids. The plastid localization of 67 of them was experimentally confirmed by proteomic approaches according to AT_CHLORO. Roughly 45–55% of all genes in the module encode components that have predicted cellular destinations other than plastids (i.e., cytosol, Golgi apparatus, nucleus, endoplasmic reticulum, plasma membrane and mitochondria). Since prediction is not 100% precise and dual localization may also occur the results must be taken with care. Nevertheless, the data suggest that RB signals affect also other parts of the cell besides plastids. We noted that 80% of the genes with enhanced expression upon blocked chloroplast biogenesis belong to the group of genes encoding non-plastid localized components. Vice versa, 94% of the genes encoding plastid localized components displayed a low expression upon blocked chloroplast biogenesis indicating that the genes within the identified core module are dominated by two opposing expression modes that are largely associated with the potential localization of the affected gene products. In order to test whether or not the opposing expression modes are also associated to specific functions we analyzed the available information in a number of databases and ordered the genes of the core module into functional subsets with at minimum four genes (Figure 2A–F) and analyzed the corresponding expression profiles (see below for more details). 23 genes remained uncategorized and are given in the supplement (Supplemental Figure S2)

3.4. Photosynthesis

In current literature photosynthesis-associated nuclear genes (PhANGs) are referenced as the "classical" target of biogenic retrograde signals [24,35]. Our meta-analysis indicates that PhANGs represent only a part of the gene groups being affected when chloroplast biogenesis is blocked (Figure 1C). Nevertheless, this group appears to be the one exhibiting the highest homogeneity in their mode of expression change among all groups of the module (Figure 2A–G) displaying a very low expression when chloroplast biogenesis is blocked while being highly induced upon etioplast-to-chloroplast conversion (WT-light vs. WT-dark). Interestingly, we identified only genes for components of the photosystem I (PSI) complex and for peripheral parts of photosystem II (PSII) including the light-harvesting antenna and the water splitting complex. Genes for subunits of the ATP-synthase, the $Cytb_6f$-complex and the NAD (P) H-dehydrogenase-like (NDH) complex as well as for enzymes of the Calvin-Benson cycle (with the phosphoribulokinase as only exception) were not identified. This implies a model in which RB signals modulate gene expression of PhANGs in a subset-specific way rather than a common overall control. It appears that especially components of the chlorophyll containing complexes respond to RB signals. Concomitant with this we observed a significant impact of RB signals on genes for key enzymes of Chl and carotenoid biosynthesis suggesting that RB signals may coordinate the syntheses of pigments and pigment-containing complexes.

A) Photosynthesis

Photosynthesis - Light reaction					
Locus	Description	WT LIN vs. WT	WT NF vs. WT	*pap7-1* vs. WT	WT light vs. WT dark
AT3G27690	LHCB2.3 (PSII light-harvesting complex protein 2.3) [a,b]	-5.10	-5.32	-1.87	2.63
AT1G76100	PETE-1 (Plastocyanin 1)[a,b]	-4.55	-5.16	-1.04	2.25
AT4G05180	PSBQ-2 (PSII subunit Q2)[a,b]	-4.55	-5.08	-1.56	2.77
AT1G52230	PSAH-2 (PSI subunit H2)	-4.14	-4.06	-0.88	1.63
AT3G54890	LHCA1 (PSI light harvesting complex protein 1)[a,b]	-3.49	-3.22	-1.80	2.27
AT1G44575	NPQ4 (Non photochemical quenching 4) [a,b]	-3.40	-3.45	-0.44	1.36
AT1G19150	LHCA6 (PSI light harvesting complex protein 6)[a,b]	-3.31	-4.58	-0.61	1.35
AT1G08380	PSAO (PSI subunit O)[a,b]	-3.13	-3.49	-1.49	2.93
AT4G02770	PSAD-1 (PSI subunit D1)[a,b]	-3.07	-3.62	-1.12	1.95
AT4G28750	PSAE-1 (PSI subunit E1)[a,b]	-2.98	-3.76	-0.61	1.48
AT3G08940	LHCB4.2 (PSII light harvesting complex protein 4.2)[a,b]	-2.87	-4.36	-1.06	1.65
AT1G30380	PSAK (PSI subunit K)[a,b]	-2.68	-3.31	-0.27	1.15
AT1G55670	PSAG (PSI subunit G)[a,b]	-2.56	-2.78	-0.90	1.78
AT4G15510	PsbP family protein [a,b]	-2.30	-1.08	-0.56	1.41
AT4G10340	LHCB5 (PSII light harvesting complex protein 5)[a,b]	-2.27	-2.34	-0.28	1.06

Photosynthesis- Connected reactions					
Locus	Description	WT LIN vs. WT	WT NF vs. WT	*pap7-1* vs. WT	WT light vs. WT dark
AT1G32060	PRK (Phosphoribulokinase)[a,b]	-2.82	-2.49	-0.78	1.19
AT5G52780	PAM68-LIKE. (NAD(P)H dehydro-genase complex assembly factor)[a]	-2.28	-2.41	-0.07	1.07
AT2G35370	GDCH (Glycine decarboxylase complex H)[a,b]	-2.05	-2.89	-0.41	1.27
AT5G48790	LOW PSII ACCUMULATION protein[a]	-2.02	-1.75	-0.57	1.53

Photosynthesis - Pigment synthesis					
Locus	Description	WT LIN vs. WT	WT NF vs. WT	*pap7-1* vs. WT	WT light vs. WT dark
AT4G27440	POR B (Protochlorophyllide oxidoreductase B)[a,b]	-4.52	-3.71	-1.24	-1.87
AT1G03630	POR C (Protochlorophyllide oxidoreductase C)[a,b]	-3.50	-3.05	-1.19	1.36
AT5G17230	PSY (Phytoene synthase)[a,b]	-2.08	-1.84	-0.60	1.30
AT1G44446	CAO / CH1 (Chlorina 1)[a,b]	-1.98	-1.72	-1.08	2.05

Figure 2. *Cont.*

B) Carbohydrate metabolism and transport

Carbohydrate metabolism					
Locus	Description	WT Lin vs. WT	WT NF vs. WT	*pap7-1* vs. WT	WT light vs. WT dark
AT4G10120	ATSPS4F (Sucrose phosphate synthase 4F)[a,b]	-5.17	-4.05	-1.75	2.55
AT5G24420	PGL5 (Probable 6-phosphogluconolactonase 5)[a,b]	-4.82	-1.93	-1.42	2.20
AT1G32900	GBSS1 (Granule-bound starch synthase)[a,b]	-3.29	-3.11	-1.55	1.43
AT3G01510	LSF1 (Phosphoglucan phosphatase)[a,b]	-2.91	-1.57	-0.53	1.05
AT5G35790	G6PD1 (Glucose-6-phosphate 1-dehydrogenase 1)[a,b]	-2.33	-1.60	-0.22	1.15
AT3G10940	LSF2 (Phosphoglucan phosphatase)[a,b]	-2.11	-1.69	-1.20	1.74
AT1G03310	ATBE2/ ISA2 (Isoamylase 2)[a,b]	-1.79	-1.43	-0.55	1.06
AT5G17520	RCP1 (Root cap 1)[a,b]	-1.55	-1.32	-0.71	1.03
AT2G47180	GOLS1 (Galactinol synthase 1)[a,b]	2.16	2.43	1.38	-1.12

Transport					
Locus	Description	WT Lin vs. WT	WT NF vs. WT	*pap7-1* vs. WT	WT light vs. WT dark
AT3G01550	PPT2 (Phosphoenolpyruvate (PEP)/Phosphate translocator 2)	-4.55	-2.37	-1.86	4.81
AT1G32080	PLGG1 (ABA-drought responses/Plastidial glycolate/ glycerate translocator 1)[a,b]	-2.39	-2.43	-0.84	1.81
AT5G12860	DiT1 (Dicarboxylate transporter 1)[a,b]	-2.00	-1.72	-0.77	1.28
AT5G42070	Unknown protein[a,b]/TOM9-2[b]	-1.98	-1.91	-0.84	1.43
AT4G13800	ENOR3L3/NIPA2 (Probable magnesium transporter)[a,b]	-1.72	-1.78	-1.62	1.52
AT5G37360	Ammonium transporter 1-like protein[a]	-1.52	-2.09	-1.12	2.28
AT1G01790	KEA1 (K+ efflux antiporter 1)[a,b]	-1.39	-1.63	-0.66	1.34
AT3G02850	SKOR (Stelar K+ outward rectifier)[a,b]	-1.21	-1.19	-0.34	-0.75
AT1G32450	NRT1.5 (Nitrate transporter 1.5)[a,b]	-1.21	-1.29	-1.01	-0.50
AT5G10180	SULTR2;1/ AST68 (Sulfate transporter 68)	-1.02	-1.35	-1.76	-0.04
AT2G04050	MATE efflux family protein[a]/ DTX3 (Detoxification 3)[b]	1.96	4.25	1.10	0.51
AT3G21090	ABC transporter family protein[a,b]	2.42	1.35	1.45	-3.26

Figure 2. *Cont.*

C) Redox regulation

Locus	Description	WT Lin vs. WT	WT NF vs. WT	*pap7-1* vs. WT	WT light vs. WT dark
	Redox regulation				
AT1G77490	TAPX (Thylakoid ascorbate peroxidase)[a,b]	-2.88	-2.65	-0.52	1.38
AT4G10000	Thioredoxin family protein [a]	-2.44	-1.11	-0.59	1.21
AT4G04610	APR1 (APS reductase 1)	-2.42	-1.66	-2.22	2.09
AT1G09500	CAD (Cinnamyl-alcohol dehydrogenase family)[a]/Phytochrome [b]	-2.37	-2.64	1.12	0.45
AT3G02730	TRXF1 (Thioredoxin f-type 1)[a,b]	-2.33	-2.64	-1.12	2.77
AT5G65840	Thioredoxin superfamily protein[a,b]	-2.23	-1.37	[illegible]	1.60
AT2G29320	NAD(P)-binding Rossmann-fold superfamily protein[a,b]	-1.74	-1.62	0.14	0.91
AT3G26060	ATPRX Q (Peroxiredoxin Q)[a,b]	-1.58	-2.09	-0.51	1.27
AT5G16710	DHAR3 (Dehydroascorbate reductase 1)[a,b]	-1.37	-1.13	-0.49	1.46
AT5G21100	AAO (L-ascorbate oxidase)[a,b]	-1.15	-1.25	-1.36	1.11
AT5G50160	FRO8 (Ferric reduction oxidase 8)[a,b]	-1.15	-1.37	-1.44	1.70
AT5G06690	WCRKC1 (WCRKC thioredoxin 1)[a,b]	-1.02	-1.65	1.20	-0.11
AT5G56090	COX15 (Cytochrome c oxidase 15)[a,b]	1.14	1.79	0.31	0.84
AT1G03850	GRXS13 (Glutaredoxin family protein)[a,b]	1.63	1.59	1.33	-1.51
AT4G15690	GRXS5 (Monothiol glutaredoxin-S5)[a,b]	1.84	2.44	0.88	0.34
AT4G15700	GRXS3/ROXY11 (Glutaredoxin family protein 3)[a,b]	2.21	2.21	1.79	3.12
AT5G64120	PRX71 (Peroxidase 71)[a,b]	2.24	1.68	1.52	-0.39
AT4G15660	GRXS8/ROXY15 (Glutaredoxin family protein 8)[a,b]	2.71	1.79	1.65	1.87

Figure 2. *Cont.*

D) Development

Organelle development						
Locus	Description	WT Lin vs. WT	WT NF vs. WT	*pap7-1* vs. WT	WT light vs. WT dark
AT1G52220	CURT1C (Curvature thylakoid 1C)[a,b]	-2.40	-1.98	-1.67	2.57
AT1G29395	COR414-TM1 (Cold regulated 414 thylakoid membrane 1)[a,b]	-2.04	-2.93	-1.24	-1.23
AT4G01150	CURT1A (Curvature thylakoid 1A)[a,b]	-1.93	-1.66	-0.58	1.14
AT3G52750	FTSZ2-2 (plastidial division protein)[a,b]	-1.92	-1.35	-0.63	1.15
AT2G46820	CURT1B (Curvature thylakoid 1B)[a,b]	-1.53	-2.04	-0.33	1.24
AT3G10420	SPD1 (Seedling plastid development 1)[a,b]	-1.49	-1.34	-1.39	1.06
AT2G45740	PEX11D (Peroxin 11D)[a,b]	-1.07	-1.61	0.32	0.84
AT1G49240	ACT8 (Actin 8)[a,b]	1.96	1.14	0.92	-1.45

Plant development						
Locus	Description	WT Lin vs. WT	WT NF vs. WT	*pap7-1* vs. WT dark	WT light vs. WT dark
AT3G28270	AFL1 (Endomembrane and plasma membrane protein)[a]	-5.63	-8.32	-1.20	3.26
AT1G23010	LPR1 (Low Phosphate Root1)/Multicopper oxidase 1[a,b]	2.12	1.13	0.48	0.62
AT5G20520	WAV2 (Wavy growth 2)[a,b]	2.28	1.28	1.57	-1.03
AT3G53230	Cell division cycle protein 48[a,b]	2.47	2.70	0.56	-1.46
AT5G20230	ATBCB (Blue-copper-binding protein)	4.39	1.82	4.07	-0.72
AT4G35770	SEN1 (Senescence 1)	5.29	-1.80	3.72	-3.37

Figure 2. *Cont.*

E) Transcription

	Transcription factors					
Locus	Description	WT Lin vs. WT	WT NF vs. WT	*pap7-1* vs. WT	WT light vs. WT dark
AT1G34310	ARF12 (Auxin response factor 12)[a,b]	-1.65	-2.80	-1.16	1.03
AT2G20570	GLK1 (Golden-like 1) [a,b]	-4.11	-1.93	-2.04	2.08
AT1G73870	COL7 (CONSTANS-like 7B-box type zinc finger protein)[a,b]	-3.59	-5.55	-1.12	1.02
AT5G07690	MYB29 (MYB domain protein 29)[a,b]	-2.35	-1.59	-2.32	5.59
AT3G56400	WRKY70 (WRKY DNA-binding protein 70)[a,b]	-1.85	-1.36	-0.57	-0.43
AT2G38090	MYB family transcription factor[a,b]	-1.02	-1.24	-0.24	-0.81
AT5G37260	RVE2 (MYB domain protein Reveille 2)[a,b]	1.16	-1.13	1.64	-1.14
AT4G26150	CGA1 (Cytokinin-responsive GATA factor 1) /GATA22[a,b]	2.12	1.49	1.96	3.19
AT3G61630	CRF6 (Cytokinin response factor 6)[a,b]	3.84	2.73	0.73	-1.12

	Transcription and RNA metabolism					
Locus	Description	WT Lin vs. WT	WT NF vs. WT	*pap7-1* vs. WT	WT light vs. WT dark
AT2G34620	mTERF10 (Mitochondrial transcription termination factor-related)[a,b]	-6.08	-3.68	-1.69	3.19
AT5G66520	CREF7 (Chloroplast RNA Editing Factor 7)[a]	-2.91	-1.68	0.47	0.84
AT4G18740	Transcription Rho termination factor[a,b]	-2.15	-1.48	0.16	0.91
AT4G25170	DNA-directed RNA polymerase subunit beta[b]	-1.69	-2.29	0.54	-1.08
AT1G06720	P-loop containing nucleoside triphosphate hydrolases superfamily protein[a,b]	1.33	2.12	0.27	0.79
AT4G15770	RNA binding / protein binding[a].60S ribosome subunit biogenesis protein NIP7 homolog[b]	1.76	1.53	0.25	0.80
AT5G45140	NRPC2 (Nuclear RNA polymerase C2)[a,b]	1.82	1.27	0.41	0.84
AT1G29940	NRPA2 (Nuclear RNA polymerase A2)[a,b]	1.85	1.36	0.29	0.73
AT1G17960	Threonyl-tRNA synthetase. putative[a,b]	3.87	3.19	1.03	0.37
AT1G19530	DNA polymerase epsilon catalytic subunit a [a,b]	3.42	2.61	2.83	-3.28

Figure 2. *Cont.*

F) Proteins and stress

Protein folding and proteostasis					
Locus	Description	WT Lin vs. WT	WT NF vs. WT	pap7-1 vs. WT	WT light vs. WT dark
AT4G39710	FKBP-type peptidyl-prolyl cis-trans isomerase family protein[a,b]	-4.22	-3.24	-0.71	1.37
AT2G27420	Cysteine proteinase, putative [a]	-4.07	-4.72	-1.95	3.55
AT5G13410	Immunophilin / FKBP-type peptidyl-prolyl cis-trans isomerase family protein[a,b]	-2.87	-1.83	-0.49	1.04
AT5G62140	ATP-dependent CLP protease ATP-binding subunit[a,b]/Galectin domain-containing protein[b]	-1.81	-2.06	-0.92	1.61
AT5G07020	MPH1 (Maintenance of PSII under high light 1)[a,b]	-1.34	-2.22	-0.30	1.18
AT3G07770	HSP89.1 (Heat shock protein 89.1)[a,b]	1.35	1.31	-0.28	1.17
AT2G17440	PIRL5 (Plant intracellular RAS-group-related LRR protein 5) [a,b]	1.55	1.40	1.29	-1.62
AT3G23990	HSP60 (Heat shock protein 60)	1.90	2.27	-0.16	1.03
AT5G09590	MTHSC70-2 (Mitochondrial HSP70 2)[a,b]	2.45	2.91	-0.20	1.15
AT5G60250	Zinc finger (C3HC4-type ring finger family protein)[a,b]	3.17	1.88	1.14	0.75
AT1G76410	ATL8 (RING/U-BOX superfamily protein)[a,b]	4.00	1.13	1.21	-1.36

Stress					
Locus	Description	WT Lin vs. WT	WT NF vs. WT	pap7-1 vs. WT	WT light vs. WT dark
AT4G23290	Cysteine-rich receptor-like protein kinase 21 CRK21 [a,b]	-4.97	-3.06	-2.99	4.53
AT2G35960	NHL12 (HIN1-LIKE 12)[a,b]	-2.57	-1.44	-0.35	1.19
AT2G42750	DNAJ heat shock N-terminal domain-containing protein[a]	-1.66	-1.64	-1.20	2.56
AT5G57345	OxR (Abiotic stress induced protein)[a,b]	-1.55	-1.56	-1.28	2.77
AT1G31580	ECS1 (cell wall protein)[a,b]	-1.21	-3.13	-1.66	6.17
AT5G64460	Phosphoglycerate mutase-like protein[a,b]	-1.15	-1.16	0.06	0.95
AT1G77120	ADH1 (Alcohol dehydrogenase 1)[a,b]	1.62	1.54	-0.80	1.08
AT5G24280	GMI1 (Gamma irradiation and mytomycin c induced 1)[a,b]	3.10	1.69	0.73	0.64
AT1G19610	PDF1.4 (Plant defensin family)[a,b]	3.53	2.46	3.00	-1.53

Figure 2. *Cont.*

G) Lipids and hormones

Lipid metabolism						
Locus	Description	WT Lin vs. WT	WT NF vs. WT	*pap7-1* vs. WT	WT light vs. WT dark
AT5G48490	DEG15. DIR1-Like (Lipid transfer protein family protein) [a]	-6.63	-3.85	-1.92	2.66
AT3G53980	Lipid transfer family protein[a,b]	1.01	-3.32	0.59	-1.46
AT1G67860	Lipase GDSL domain-containing protein[b]	1.83	-1.57	-0.97	-1.46
AT3G21720	ICL (Isocitrate lyase)[a,b]	2.31	3.68	1.90	3.26
AT2G37870	Lipid transfer family protein[a,b]	2.53	-2.28	0.39	1.15

Hormone synthesis and signalling						
Locus	Description	WT Lin vs. WT	WT NF vs. WT	*pap7-1* vs. WT	WT light vs. WT dark
AT3G45140	LOX2 (Lipoxygenase 2)[a,b]	-4.84	-5.00	-2.01	6.11
AT4G12980	DEG18 (Auxin-responsive family protein gene)[a]. Cytochrome b561 and DOMON domain-containing protein[b]	-2.77	-1.36	-1.23	1.11
AT4G23600	CORI3 (Coronatine induced 1)[a,b]	-2.52	-2.39	-1.36	3.02
AT5G45820	CIPK20 (Cbl-interacting protein kinase 20)[a,b]	-2.37	-1.60	-1.31	3.58
AT5G42650	AOS (Allene oxide synthase)/ CYP74A (Cytochrome P450 74A)/ DDE2 (Delayed dehiscence 2) [a,b]	-1.87	-2.65	-2.20	1.44
AT1G12010	ACO3 (1-aminocyclopropane-1-carboxylate oxidase. putative)[a,b]	1.18	2.12	1.27	-1.12
AT5G37990	CIMT1 (SABATH gene family)[a,b]	1.34	-3.16	-0.33	1.37
AT1G15670	KMD2 (Kiss me deadly)[a,b]	1.49	1.37	0.17	-1.15
AT5G14920	GASA14 (Gibberellin-regulated family protein 14)[a,b]	1.94	-1.96	0.38	-1.22

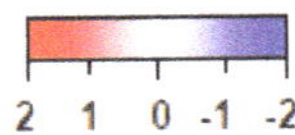

Figure 2. Expression patterns of selected functional groups within the identified core module. Micro-array based expression data of light-grown seedlings treated with LIN or NF are given in comparison to wild type (WT LIN vs. WT and WT NF vs. WT, respectively). Expression data from the *pap7-1* mutant are given in comparison to light-grown WT (*pap7-1* vs. WT light), and dark grown WT (*pap7-1* vs. WT dark). As general control expression data of light-grown WT in relation to dark-grown WT are given. All data represent log2-fold expression changes and are supported by colour code indicated in the bottom right corner of figure G. Gene identities (Locus) and respective encoded proteins (Description) are given in columns to the left. Functional groups are indicated on top of each table and are arranged in functionally related subsets. (**A**) Photosynthesis, (**B**) carbohydrate metabolism and transport, (**C**) redox regulation, (**D**) development, (**E**) transcription, (**F**) proteins and stress, (**G**) lipids and hormones. Genes encoding proteins with predicted or proven plastid localization are highlighted in light-green. For details see Supplemental Table S1.

3.5. Carbohydrate Metabolism and Transport

The direct products of photosynthesis are ATP and NADPH that are primarily used in the Calvin cycle to generate carbohydrates. However, they are also used in many other metabolic pathways that reside either partially or entirely in the plastid. Inhibition or down-regulation of photosynthesis, therefore, has a profound impact on the overall metabolic capacities of plastids. This is reflected in the effect of RB signals on the expression of numerous genes encoding plastid and cytosolic enzymes involved in carbohydrate metabolism that exhibit the same expression profile as the photosynthesis genes. Only the gene *GOLS1* (encoding the galactinol-synthase 1) exhibits an opposite expression profile. *GOLS1* has been proposed as a negative regulator of seed germination [36] as well as a gene responsive to various stressors. It is, thus, conceivable that RB signals from blocked chloroplast biogenesis indicate an early developmental plastid stage which is similar to that of proplastids in seeds or etioplasts of dark grown seedlings. This signal potentially arrests the gene expression programme that normally occurs when a seedling is exposed to light. In line with this assumption is the observation that a number of genes for transporter proteins required for the metabolic exchange across the chloroplast envelope show low expression. This accounts also for some transporters outside the plastid that are required to establish metabolic pipelines that end in a plastid (e.g., such as the nitrate transporter).

3.6. Redox Regulation

Many enzymatic reactions in chloroplasts are regulated via the redox state of the corresponding proteins. This control is exerted by a number of redox mediators such as thioredoxins that derive their reduction power from photosynthesis [37]. Many of the genes in this subset exhibit an expression profile similar to those within the photosynthesis module. The same is observed for genes encoding components of the antioxidant network (such as the gene for the dehydroascorbate reductase). All of them encode plastid localized proteins. Different expression profiles have been detected for several genes for cytosolic glutaredoxins that show strong accumulation. Many of these glutaredoxins are known to be involved in pathogen defense or stress responses [38,39]. Also other subsets contain components of the pathogen and stress response system suggesting that it is a major functional target of RB signals.

3.7. Development

As novel highly interesting targets for RB signals we identified three of the four genes encoding CURT (Curvature thylakoid) proteins within the core module. These proteins are required for the formation of the grana margins in thylakoid membrane system [40,41] and, therefore, absolutely essential for the build-up of the photosynthetic apparatus. Another novel target gene encodes SPD1 (seedling plastid development 1), a components required for eoplast de-differentiation and, therefore, being involved in very early developmental processes of chloroplast biogenesis [42]. All exhibit an expression profile corresponding to the photosynthesis subset. The same expression pattern was observed for the gene for peroxin 11, a component required for the build-up of peroxisomes [43]. Peroxisomes are functionally tightly coupled to chloroplasts as part of the photorespiratory pathway, but also in the synthesis of jasmonates in pathogen and wound response. Interestingly, an opposite expression pattern was observed for the gene encoding ACT8 (actin 8), a protein presumably involved in chloroplast movement or positioning suggesting that proper positioning may be required in early plastid biogenesis [44]. More general components involved in various aspects of plant development were also affected by RB signals; however, they exhibit expression patterns that are difficult to interpret. Only one gene in this subset encodes a plastid-localized protein. This protein is SEN1 (senescence 1), a component that is associated with senescence and strongly induced in the dark (therefore also known as dark-inducible 1) and upon phosphate starvation and biotic stresses [45]. It represents the gene with the strongest opposing expression between LIN and NF treatments (Supplemental Figure S3). It was also observed to be strongly induced by abscisic acid (ABA) [46]. Since NF is blocking carotenoid production, and hence the ABA precursors, the opposing expression pattern is likely caused by this inhibitor-specific difference.

3.8. Transcription

Special interest in retrograde signaling is paid to the identification of nuclear localized transcription factors that mediate the corresponding gene expression responses (Figure 2E). An often discussed prime candidate, ABI4 (ABA insensitive 4), has been recently shown to be likely not involved in retrograde signaling [47]. Our meta-analysis identified nine genes encoding other potential candidates that exhibit diverse expression patterns. One highly interesting candidate is GLK1 that exhibits the "photosynthesis-type" transcript accumulation. It was shown to be a major activator of PhANG expression and chloroplast biogenesis [48] and is also down-regulated at protein level in *pap8-1*, an albino mutant related to *pap7-1* [49]. The low expression of GLK1 in response to blocked chloroplast biogenesis likely accounts for a large part of the expression profiles observed here. We noticed that the partner regulator GLK2 did not appear in our gene list suggesting that it is regulated differently. GLK1 may potentially work together with another candidate, COL7, a transcription factor involved in light signaling that interacts with HY5 (Supplemental Table S1). The gene for the MYB-domain containing transcription factor MYB29 shares the "photosynthesis-type" expression pattern. It is known to be a major activator of the biosynthesis of glucosinolates, secondary metabolites involved in responses to biotic and abiotic stresses [50]. This observation aligns well with the expression patterns of MAM1 and BGL28 (encoding two enzymes required for glucosinolate biosynthesis) (Supplemental Figure S2 and Table S1). MYB29 is also involved in retrograde control from mitochondria regulating the expression of alternative oxidase 1a providing an interesting link between the two organelles [51]. Genes for three transcription factors RVE2, CGA1 and CRF6 (Reveille 2, Cytokinin-responsive GATA factor 1 and Cytokinin response factor 6) exhibit enhanced expression upon block of chloroplast biogenesis. Reveille 2, also a MYB-domain containing factor, is known to promote primary dormancy, but is repressed under illumination and imbibition by PhyB [52]. Its opposite expression in the NF sample may be an ABA-mediated effect as with SEN1 (see above). CGA1 and CRF6 are both implicated in cytokinin signaling. Cytokinin activates chloroplast biogenesis by inducing multiple target genes, among them HY5, CGA1 and CRF6. CGA1 is also known to act in an additive manner to GLK1 and both together represent a major regulatory hub of chloroplast biogenesis with GLKs likely acting down-stream of GNCs [53,54]. A recent study revealed that *pap* mutants overproduce cytokinins [55]. We regard it as likely that blocking of chloroplast biogenesis by LIN or NF induce similar responses. In sum, we identified a set of key transcription factors that is most likely involved in the mediation of RB signals during early chloroplast development and provides interesting candidates for future research in this context. Apart from directly affecting the expression of these key transcription regulators, RB signals appear to affect also other components involved in transcription or RNA metabolism both in plastid and in nucleus. A number of recent studies have identified numerous connections between retrograde signals and RNA metabolism in nucleus and plastid pointing to a potentially important regulatory level [56–59].

3.9. Proteins and Stress

Our meta-analysis was focused on the implications of RB signals on transcript accumulation and cannot account for potential effects on other gene expression levels. However, we observed significant effects of RB signals on genes encoding components involved in protein folding, proteostasis and stress responses, which imply that blocking chloroplast biogenesis affects also these functions. Recent studies revealed a major impact of plastid signals on the unfolded protein response in plastids, endoplasmic reticulum and cytosol as well as on protein accumulation [60–62]. This includes the action of heat shock proteins and likely other functionally related proteins such as immunophilins. The GLK1 transcription factor has been recently shown to be controlled at the protein level by RB signals [63] providing an important example. This aspect of retrograde signaling certainly will likely expand a lot in future research.

3.10. Lipids and Hormones

Lipids are major constituents of membranes and are absolutely essential for the build-up of chloroplast. Although blocking of chloroplast biogenesis results in plastids without thylakoid formation we did not find any impact on genes encoding proteins involved in plastid lipid metabolism, but only for non-plastid components. The corresponding gene expression profiles are very complex and difficult to interpret and more analyses will be necessary to understand and to explain the RB impact on them. We also found an impact of RB signals on multiple genes involved in hormone biosynthesis and/or signaling. We identified already in other functional subsets a number of candidate genes that are associated with synthesis and/or the action of several hormones (see above). ABA appears to be an interceptive signal in a few cases (Supplemental Figure S3), but likely is not a major contributor to RB signaling in chloroplast biogenesis since NF treatment results in the same expression pattern as in the LIN and *pap7-1* profiles.

Interestingly, we found a strong impact on two genes encoding key enzymes of allene oxide biosynthesis (lipoxygenase and allene oxide synthase) that produce the precursor for peroxisomal jasmonic acid production, the oxophytodienoic acid (OPDA) [64]. A distinct role of OPDA in retrograde signaling has been not yet reported, but the molecule represents a likely candidate for a metabolite signal since it leaves the plastid for further metabolic processing. The impact on the allene oxide pathway probably is connected to the missing allene oxide precursor molecule linolenic acid that usually originates from the thylakoid lipids and which are not present in the arrested plastid [64]. A number of other genes encoding components involved in or connected to plant defense to biotic and abiotic stressors or peroxisome biogenesis were identified in this study (see above) supporting the view that a block of chloroplast biogenesis generates a situation of severe stress that is not only based on the missing photosynthetic function, but also caused by a dysregulation in the build-up of the plant defense system. Besides, most other genes identified in the "Hormones" subset encode proteins with non-plastidial locations demonstrating the broad impact of RB signals on the hormone-signaling network.

4. Conclusions

The core module responsive to RB signals identified in this study is different from retrograde controlled gene modules identified in earlier studies [65–67], most likely because these studies included conditions in which also retrograde operational (RO) signals are active (i.e., in the presence of fully developed chloroplasts). All three conditions analyzed in this study did not include the action of RO signals. The molecular targets causing the arrest in chloroplast biogenesis were different in the three conditions (cf. introduction), but the affected processes are not independent from each other and are interlinked by negative feedback loops affecting the generation of components of the transcriptional (rpo subunits) and translational (ribosomal components) machineries (Figure 3). Because of these connections and the high similarity in the three expression profiles we regard it as very likely that all three conditions generate signals that feed into the same signaling pathway(s). This signaling pathway targets mainly genes for photosynthesis or processes coupled to photosynthesis. Typically such genes are up-regulated by illumination and RB signals appear to intercept this light-dependent activation. A much smaller group of genes displayed the opposite expression profile indicating that RB signals can be of positive or negative effect. This master expression switch accounts for most genes in the core module. Only a few genes displayed more complex patterns that suggest the involvement of additional regulatory signals in their expression (such as ABA mediated signals). It must be noted that our meta-analysis describes only gene expression changes. Thus, it cannot draw conclusions on the positive or negative nature of RB signals and cannot distinguish between missing activation or active repression (or vice versa) [21,68]. It also cannot explain whether the listed genes are direct or indirect targets. We regard it as likely that most genes in this module are controlled by just a few regulators primarily targeted by transcription factors responding to the RB signal(s) (Figure 2E). This would explain the homogeneous expression profiles of so many different genes. Novel prime candidates for the mediation of the RB signal(s) towards such primary regulators are dually localized PEP-associated proteins (PAPs) that have been shown to be essential for chloroplast biogenesis and formation of late photobodies during early steps of photomorphogenesis [49,69,70]. They belong to a group of proteins that appear to move via the plastid towards the nucleus representing genuine retrograde signals [71]. The core module identified here provides novel insights into the targets of RB signals and may serve as base for more detailed working models in future studies. This will include the determination of the accumulation of the corresponding proteins.

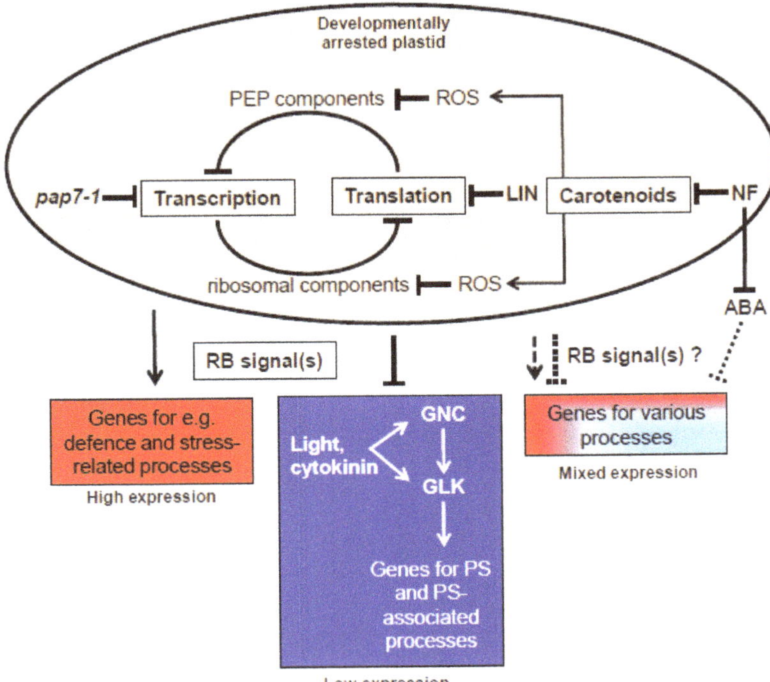

Figure 3. Working hypothesis for the action of retrograde biogenic signals. Oval on top represents a plastid that is arrested in its development either genetically (*pap7-1* mutant) or chemically (LIN or NF). The molecular inhibitory effects are connected by negative feedback loops resulting in a uniform retrograde biogenic (RB) signal in all three cases. Major target is the group of photosynthesis and photosynthesis-associated genes that normally are activated during chloroplast biogenesis by light and cytokinins with GNC and GLK transcription factors as major regulatory hubs. RB signals intercept into this activation resulting in low expression of these genes. RB signals may act also oppositely by promoting the expression of a small set of genes, many of them are involved in or related to defence responses to biotic and abiotic stresses. Largely not understood is the action of RB signals on genes with mixed expression. Likely additional regulatory factors may play a role such as the suppression of abscisic acid (ABA) formation upon NF treatment. Arrows represent conceptual positive influences, bars represent conceptual negative influences.

Supplementary Materials: The following are available online at https://www.mdpi.com/2223-7747/10/2/296/s1, Figure S1: Comparison of RB-affected genes with direct HY5 target genes, Figure S2: Genes in the core module encoding proteins not related to one of the major functional groups, Figure S3: Genes with opposite expression profiles in the LIN and NF data sets, Table S1: Analysis of 152 genes responsive to retrograde biogenic signals.

Author Contributions: Conceptualization, T.P.; Formal analysis, B.G., C.C., T.P.; Data curation, C.C., T.P.; Writing, B.G., C.C., T.P.; Supervision, T.P. All authors have read and agreed to the published version of the manuscript.

Funding: This research received no external funding.

Institutional Review Board Statement: Not applicable.

Informed Consent Statement: Not applicable.

Data Availability Statement: Original data sources are described in the Methods section.

Conflicts of Interest: The authors declare no conflict of interest.

References

1. Jarvis, P.; López-Juez, E. Biogenesis and homeostasis of chloroplasts and other plastids. *Nat. Rev. Mol. Cell Biol.* **2013**, *14*, 787. [CrossRef]
2. Liebers, M.; Grubler, B.; Chevalier, F.; Lerbs-Mache, S.; Merendino, L.; Blanvillain, R.; Pfannschmidt, T. Regulatory shifts in plastid transcription play a key role in morphological conversions of plastids during plant development. *Front. Plant. Sci.* **2017**, *8*, 23. [CrossRef]
3. Solymosi, K.; Schoefs, B. Etioplast and etio-chloroplast formation under natural conditions: The dark side of chlorophyll biosynthesis in angiosperms. *Photosynth. Res.* **2010**, *105*, 143–166. [CrossRef] [PubMed]
4. Dubreuil, C.; Jin, X.; Barajas-Lopez, J.D.; Hewitt, T.C.; Tanz, S.K.; Dobrenel, T.; Schroder, W.P.; Hanson, J.; Pesquet, E.; Gronlund, A.; et al. Establishment of photosynthesis through chloroplast development is controlled by two distinct regulatory phases. *Plant. Physiol.* **2018**, *176*, 1199–1214. [CrossRef] [PubMed]
5. Armarego-Marriott, T.; Kowalewska, L.; Burgos, A.; Fischer, A.; Thiele, W.; Erban, A.; Strand, D.; Kahlau, S.; Hertle, A.; Kopka, J.; et al. Highly resolved systems biology to dissect the etioplast-to-chloroplast transition in tobacco leaves. *Plant. Physiol.* **2019**, *180*, 654–681. [CrossRef] [PubMed]
6. Armarego-Marriott, T.; Sandoval-Ibanez, O.; Kowalewska, L. Beyond the darkness: Recent lessons from etiolation and de-etiolation studies. *J. Exp. Bot.* **2020**, *71*, 1215–1225. [CrossRef]
7. Pogson, B.J.; Ganguly, D.; Albrecht-Borth, V. Insights into chloroplast biogenesis and development. *Biochim. Biophys. Acta* **2015**, *1847*, 1017–1024. [CrossRef] [PubMed]
8. Yoo, C.Y.; Han, S.; Chen, M. Nucleus-to-plastid phytochrome signalling in controlling chloroplast biogenesis. *Ann. Plant. Rev.* **2020**, *3*, 251–280.
9. Deng, X.W.; Matsui, M.; Wei, N.; Wagner, D.; Chu, A.M.; Feldmann, K.A.; Quail, P.H. COP1, an Arabidopsis regulatory gene, encodes a protein with both a zinc-binding motif and a G beta homologous domain. *Cell* **1992**, *71*, 791–801. [CrossRef]
10. Pfalz, J.; Pfannschmidt, T. Essential nucleoid proteins in early chloroplast development. *Trends Plant. Sci.* **2013**, *18*, 186–194. [CrossRef]
11. Borner, T.; Aleynikova, A.Y.; Zubo, Y.O.; Kusnetsov, V.V. Chloroplast RNA polymerases: Role in chloroplast biogenesis. *Biochim. Biophys. Acta* **2015**, *1847*, 761–769. [CrossRef]
12. Chan, K.X.; Phua, S.Y.; Crisp, P.; McQuinn, R.; Pogson, B.J. Learning the languages of the chloroplast: Retrograde signaling and beyond. *Annu. Rev. Plant. Biol.* **2016**, *67*, 25–53. [CrossRef]
13. De Souza, A.; Wang, J.Z.; Dehesh, K. Retrograde Signals: Integrators of interorganellar communication and orchestrators of plant development. *Annu. Rev. Plant. Biol.* **2016**, *68*, 85–108. [CrossRef]
14. Grimm, B.; Dehesh, K.; Zhang, L.; Leister, D. Intracellular communication. *Mol. Plant.* **2014**, *7*, 1071–1074. [CrossRef]
15. Chi, W.; Sun, X.; Zhang, L. Intracellular signaling from plastid to nucleus. *Annu. Rev. Plant. Biol.* **2013**, *64*, 559–582. [CrossRef]
16. Inaba, T.; Yazu, F.; Ito-Inaba, Y.; Kakizaki, T.; Nakayama, K. Retrograde signaling pathway from plastid to nucleus. *Int. Rev. Cell Mol. Biol.* **2011**, *290*, 167–204. [PubMed]
17. Pfannschmidt, T.; Terry, M.J.; Van Aken, O.; Quiros, P.M. Retrograde signals from endosymbiotic organelles: A common control principle in eukaryotic cells. *Philos. Trans. R. Soc. Lond. B Biol. Sci.* **2020**, *375*, 20190396. [CrossRef] [PubMed]
18. Pogson, B.J.; Woo, N.S.; Forster, B.; Small, I.D. Plastid signalling to the nucleus and beyond. *Trends Plant. Sci.* **2008**, *13*, 602–609. [CrossRef] [PubMed]
19. Bradbeer, J.W.; Atkinson, Y.E.; Börner, T.; Hagemann, R. Cytoplasmic synthesis of plastid polypeptides may be controlled by plastid synthesized RNA. *Nature* **1979**, *279*, 816–817. [CrossRef]
20. Oelmuller, R.; Levitan, I.; Bergfeld, R.; Rajasekhar, V.K.; Mohr, H. Expression of nuclear genes as affected by treatments acting on the plastids. *Planta* **1986**, *168*, 482–492. [CrossRef]

21. Pfannschmidt, T. Plastidial retrograde signalling—A true "plastid factor" or just metabolite signatures? *Trends Plant. Sci.* **2010**, *15*, 427–435. [CrossRef]
22. Hernandez-Verdeja, T.; Strand, A. Retrograde signals navigate the path to chloroplast development. *Plant. Physiol.* **2018**, *176*, 967–976. [CrossRef]
23. Susek, R.E.; Ausubel, F.M.; Chory, J. Signal transduction mutants of Arabidopsis uncouple nuclear CAB and RBCS gene expression from chloroplast development. *Cell* **1993**, *74*, 787–799. [CrossRef]
24. Sullivan, J.A.; Gray, J.C. Plastid translation is required for the expression of nuclear photosynthesis genes in the dark and in roots of the pea lip1 mutant. *Plant. Cell* **1999**, *11*, 901–910. [PubMed]
25. Grubler, B.; Merendino, L.; Twardziok, S.O.; Mininno, M.; Allorent, G.; Chevalier, F.; Liebers, M.; Blanvillain, R.; Mayer, K.; Lerbs-Mache, S.; et al. Light and plastid signals regulate different sets of genes in the albino mutant pap7-1. *Plant. Physiol.* **2017**, *175*, 1203–1219. [CrossRef]
26. Page, M.T.; McCormac, A.C.; Smith, A.G.; Terry, M.J. Singlet oxygen initiates a plastid signal controlling photosynthetic gene expression. *New Phytol.* **2017**, *213*, 1168–1180. [CrossRef]
27. Ruckle, M.E.; Burgoon, L.D.; Lawrence, L.A.; Sinkler, C.A.; Larkin, R.M. Plastids are major regulators of light signaling in Arabidopsis. *Plant. Physiol.* **2012**, *159*, 366–390. [CrossRef] [PubMed]
28. Heberle, H.; Meirelles, G.V.; da Silva, F.R.; Telles, G.P.; Minghim, R. InteractiVenn: A web-based tool for the analysis of sets through Venn diagrams. *BMC Bioinform.* **2015**, *16*, 169. [CrossRef]
29. Usadel, B.; Poree, F.; Nagel, A.; Lohse, M.; Czedik-Eysenberg, A.; Stitt, M. A guide to using MapMan to visualize and compare Omics data in plants: A case study in the crop species, Maize. *Plant. Cell Environ.* **2009**, *32*, 1211–1229. [CrossRef] [PubMed]
30. Bruley, C.; Dupierris, V.; Salvi, D.; Rolland, N.; Ferro, M. AT_CHLORO: A chloroplast protein database dedicated to sub-plastidial localization. *Front. Plant. Sci.* **2012**, *3*, 205. [CrossRef] [PubMed]
31. Usadel, B.; Nagel, A.; Thimm, O.; Redestig, H.; Blaesing, O.E.; Palacios-Rojas, N.; Selbig, J.; Hannemann, J.; Piques, M.C.; Steinhauser, D.; et al. Extension of the visualization tool MapMan to allow statistical analysis of arrays, display of corresponding genes, and comparison with known responses. *Plant. Physiol.* **2005**, *138*, 1195–1204. [CrossRef]
32. Ruckle, M.E.; DeMarco, S.M.; Larkin, R.M. Plastid signals remodel light signaling networks and are essential for efficient chloroplast biogenesis in Arabidopsis. *Plant. Cell* **2007**, *19*, 3944–3960. [CrossRef]
33. Burko, Y.; Seluzicki, A.; Zander, M.; Pedmale, U.V.; Ecker, J.R.; Chory, J. Chimeric Activators and Repressors Define HY5 Activity and Reveal a Light-Regulated Feedback Mechanism. *Plant. Cell* **2020**, *32*, 967–983. [CrossRef]
34. Ma, L.; Li, J.; Qu, L.; Hager, J.; Chen, Z.; Zhao, H.; Deng, X.W. Light control of Arabidopsis development entails coordinated regulation of genome expression and cellular pathways. *Plant. Cell* **2001**, *13*, 2589–2607. [CrossRef] [PubMed]
35. Koussevitzky, S.; Nott, A.; Mockler, T.C.; Hong, F.; Sachetto-Martins, G.; Surpin, M.; Lim, J.; Mittler, R.; Chory, J. Signals from chloroplasts converge to regulate nuclear gene expression. *Science* **2007**, *316*, 715–719. [CrossRef]
36. Jang, J.H.; Shang, Y.; Kang, H.K.; Kim, S.Y.; Kim, B.H.; Nam, K.H. Arabidopsis galactinol synthases 1 (AtGOLS1) negatively regulates seed germination. *Plant. Sci.* **2018**, *267*, 94–101. [CrossRef]
37. Dietz, K.J.; Pfannschmidt, T. Novel regulators in photosynthetic redox control of plant metabolism and gene expression. *Plant. Physiol.* **2011**, *155*, 1477–1485. [CrossRef] [PubMed]
38. Gutsche, N.; Holtmannspotter, M.; Mass, L.; O'Donoghue, M.; Busch, A.; Lauri, A.; Schubert, V.; Zachgo, S. Conserved redox-dependent DNA binding of ROXY glutaredoxins with TGA transcription factors. *Plant. Direct.* **2017**, *1*, e00030. [CrossRef] [PubMed]
39. Ndamukong, I.; Abdallat, A.A.; Thurow, C.; Fode, B.; Zander, M.; Weigel, R.; Gatz, C. SA-inducible Arabidopsis glutaredoxin interacts with TGA factors and suppresses JA-responsive PDF1.2 transcription. *Plant. J.* **2007**, *50*, 128–139. [CrossRef]
40. Armbruster, U.; Labs, M.; Pribil, M.; Viola, S.; Xu, W.; Scharfenberg, M.; Hertle, A.P.; Rojahn, U.; Jensen, P.E.; Rappaport, F.; et al. Arabidopsis CURVATURE THYLAKOID1 proteins modify thylakoid architecture by inducing membrane curvature. *Plant. Cell* **2013**, *25*, 2661–2678. [CrossRef]
41. Pribil, M.; Labs, M.; Leister, D. Structure and dynamics of thylakoids in land plants. *J. Exp. Bot.* **2014**, *65*, 1955–1972. [CrossRef]
42. Ruppel, N.J.; Logsdon, C.A.; Whippo, C.W.; Inoue, K.; Hangarter, R.P. A mutation in Arabidopsis seedling plastid development1 affects plastid differentiation in embryo-derived tissues during seedling growth. *Plant. Physiol.* **2011**, *155*, 342–353. [CrossRef]
43. Lingard, M.J.; Gidda, S.K.; Bingham, S.; Rothstein, S.J.; Mullen, R.T.; Trelease, R.N. Arabidopsis PEROXIN11c-e, FISSION1b, and DYNAMIN-RELATED PROTEIN3A cooperate in cell cycle-associated replication of peroxisomes. *Plant. Cell* **2008**, *20*, 1567–1585. [CrossRef]
44. Sheahan, M.B.; Collings, D.A.; Rose, R.J.; McCurdy, A.D.W. ACTIN7 is required for perinuclear clustering of chloroplasts during arabidopsis protoplast culture. *Plants* **2020**, *9*, 225. [CrossRef]
45. Fernandez-Calvino, L.; Guzman-Benito, I.; Del Toro, F.J.; Donaire, L.; Castro-Sanz, A.B.; Ruiz-Ferrer, V.; Llave, C. Activation of senescence-associated Dark-inducible (DIN) genes during infection contributes to enhanced susceptibility to plant viruses. *Mol. Plant. Pathol.* **2016**, *17*, 3–15. [CrossRef] [PubMed]
46. Oh, S.A.; Lee, S.Y.; Chung, I.K.; Lee, C.H.; Nam, H.G. A senescence-associated gene of Arabidopsis thaliana is distinctively regulated during natural and artificially induced leaf senescence. *Plant. Mol. Biol.* **1996**, *30*, 739–754. [CrossRef] [PubMed]
47. Kacprzak, S.M.; Mochizuki, N.; Naranjo, B.; Xu, D.; Leister, D.; Kleine, T.; Okamoto, H.; Terry, M.J. Plastid-to-nucleus retrograde signalling during chloroplast biogenesis does not require ABI4. *Plant. Physiol.* **2019**, *179*, 18–23. [CrossRef] [PubMed]

48. Waters, M.T.; Wang, P.; Korkaric, M.; Capper, R.G.; Saunders, N.J.; Langdale, J.A. GLK transcription factors coordinate expression of the photosynthetic apparatus in Arabidopsis. *Plant. Cell* **2009**, *21*, 1109–1128. [CrossRef] [PubMed]
49. Liebers, M.; Gillet, F.X.; Israel, A.; Pounot, K.; Chambon, L.; Chieb, M.; Chevalier, F.; Ruedas, R.; Favier, A.; Gans, P.; et al. Nucleo-plastidic PAP8/pTAC6 couples chloroplast formation with photomorphogenesis. *EMBO J.* **2020**, *39*, e104941. [CrossRef] [PubMed]
50. Zuluaga, D.L.; Graham, N.S.; Klinder, A.; van Ommen Kloeke, A.E.E.; Marcotrigiano, A.R.; Wagstaff, C.; Verkerk, R.; Sonnante, G.; Aarts, M.G.M. Overexpression of the MYB29 transcription factor affects aliphatic glucosinolate synthesis in Brassica oleracea. *Plant. Mol. Biol.* **2019**, *101*, 65–79. [CrossRef]
51. Zhang, X.; Ivanova, A.; Vandepoele, K.; Radomiljac, J.; Van de Velde, J.; Berkowitz, O.; Willems, P.; Xu, Y.; Ng, S.; Van Aken, O.; et al. The transcription factor MYB29 is a regulator of ALTERNATIVE OXIDASE1a. *Plant. Physiol.* **2017**, *173*, 1824–1843. [CrossRef]
52. Jiang, Z.; Xu, G.; Jing, Y.; Tang, W.; Lin, R. Phytochrome B and REVEILLE1/2-mediated signalling controls seed dormancy and germination in Arabidopsis. *Nat. Commun.* **2016**, *7*, 12377. [CrossRef]
53. Lopez-Juez, E. Plastid biogenesis, between light and shadows. *J. Exp. Bot.* **2007**, *58*, 11–26. [CrossRef]
54. Zubo, Y.O.; Blakley, I.C.; Franco-Zorrilla, J.M.; Yamburenko, M.V.; Solano, R.; Kieber, J.J.; Loraine, A.E.; Schaller, G.E. Coordination of chloroplast development through the action of the GNC and GLK transcription factor families. *Plant. Physiol.* **2018**, *178*, 130–147. [CrossRef]
55. Andreeva, A.A.; Vankova, R.; Bychkov, I.A.; Kudryakova, N.V.; Danilova, M.N.; Lacek, J.; Pojidaeva, E.S.; Kusnetsov, V.V. Cytokinin-regulated expression of arabidopsis thaliana PAP genes and its implication for the expression of chloroplast-encoded genes. *Biomolecules* **2020**, *10*, 1658. [CrossRef] [PubMed]
56. Zhao, X.; Huang, J.; Chory, J. Unraveling the linkage between retrograde signaling and RNA metabolism in plants. *Trends Plant. Sci.* **2020**, *25*, 141–147. [CrossRef]
57. Petrillo, E.; Godoy Herz, M.A.; Fuchs, A.; Reifer, D.; Fuller, J.; Yanovsky, M.J.; Simpson, C.; Brown, J.W.; Barta, A.; Kalyna, M.; et al. A chloroplast retrograde signal regulates nuclear alternative splicing. *Science* **2014**, *344*, 427–430. [CrossRef]
58. Fang, X.; Zhao, G.; Zhang, S.; Li, Y.; Gu, H.; Li, Y.; Zhao, Q.; Qi, Y. Chloroplast-to-Nucleus Signaling Regulates MicroRNA Biogenesis in Arabidopsis. *Dev. Cell* **2019**, *48*, 371–382.e4. [CrossRef] [PubMed]
59. Zhao, X.; Huang, J.; Chory, J. GUN1 interacts with MORF2 to regulate plastid RNA editing during retrograde signaling. *Proc. Natl. Acad. Sci. USA* **2019**, *116*, 10162–10167. [CrossRef] [PubMed]
60. Llamas, E.; Pulido, P.; Rodriguez-Concepcion, M. Interference with plastome gene expression and Clp protease activity in Arabidopsis triggers a chloroplast unfolded protein response to restore protein homeostasis. *PLoS Genet.* **2017**, *13*, e1007022. [CrossRef]
61. Wu, G.Z.; Meyer, E.H.; Richter, A.S.; Schuster, M.; Ling, Q.; Schottler, M.A.; Walther, D.; Zoschke, R.; Grimm, B.; Jarvis, R.P.; et al. Control of retrograde signalling by protein import and cytosolic folding stress. *Nat. Plants* **2019**, *5*, 525–538. [CrossRef] [PubMed]
62. Beaugelin, I.; Chevalier, A.; D'Alessandro, S.; Ksas, B.; Havaux, M. Endoplasmic reticulum-mediated unfolded protein response is an integral part of singlet oxygen signalling in plants. *Plant. J.* **2020**, *102*, 1266–1280. [CrossRef]
63. Tokumaru, M.; Adachi, F.; Toda, M.; Ito-Inaba, Y.; Yazu, F.; Hirosawa, Y.; Sakakibara, Y.; Suiko, M.; Kakizaki, T.; Inaba, T. Ubiquitin-proteasome dependent regulation of the GOLDEN2-LIKE 1 transcription factor in response to plastids signals. *Plant. Physiol.* **2017**, *173*, 524–535. [CrossRef] [PubMed]
64. Maynard, D.; Groger, H.; Dierks, T.; Dietz, K.J. The function of the oxylipin 12-oxophytodienoic acid in cell signaling, stress acclimation, and development. *J. Exp. Bot.* **2018**, *69*, 5341–5354. [CrossRef]
65. Glasser, C.; Haberer, G.; Finkemeier, I.; Pfannschmidt, T.; Kleine, T.; Leister, D.; Dietz, K.J.; Hausler, R.E.; Grimm, B.; Mayer, K.F. Meta-analysis of retrograde signaling in Arabidopsis thaliana reveals a core module of genes embedded in complex cellular signaling networks. *Mol. Plant.* **2014**, *7*, 1167–1190. [CrossRef] [PubMed]
66. Leister, D.; Romani, I.; Mittermayr, L.; Paieri, F.; Fenino, E.; Kleine, T. Identification of target genes and transcription factors implicated in translation-dependent retrograde signaling in Arabidopsis. *Mol. Plant.* **2014**, *7*, 1228–1247. [CrossRef] [PubMed]
67. Leister, D.; Kleine, T. Definition of a core module for the nuclear retrograde response to altered organellar gene expression identifies GLK overexpressors as gun mutants. *Physiol. Plant.* **2016**, *157*, 297–309. [CrossRef]
68. Terry, M.J.; Smith, A.G. A model for tetrapyrrole synthesis as the primary mechanism for plastid-to-nucleus signaling during chloroplast biogenesis. *Front. Plant. Sci.* **2013**, *4*, 14. [CrossRef]
69. Chen, M.; Galvao, R.M.; Li, M.; Burger, B.; Bugea, J.; Bolado, J.; Chory, J. Arabidopsis HEMERA/pTAC12 initiates photomorphogenesis by phytochromes. *Cell* **2010**, *141*, 1230–1240. [CrossRef]
70. Yang, E.J.; Yoo, C.Y.; Liu, J.; Wang, H.; Cao, J.; Li, F.W.; Pryer, K.M.; Sun, T.P.; Weigel, D.; Zhou, P.; et al. NCP activates chloroplast transcription by controlling phytochrome-dependent dual nuclear and plastidial switches. *Nat. Commun.* **2019**, *10*, 2630. [CrossRef]
71. Krupinska, K.; Blanco, N.E.; Oetke, S.; Zottini, M. Genome communication in plants mediated by organelle-nucleus-located proteins. *Philos. Trans. R. Soc. Lond. B Biol. Sci.* **2020**, *375*, 20190397. [CrossRef] [PubMed]

Article

Mitochondrial mRNA Processing in the Chlorophyte Alga *Pediastrum duplex* and Streptophyte Alga *Chara vulgaris* Reveals an Evolutionary Branch in Mitochondrial mRNA Processing

Grayson C. R. Proulex [1,†], Marcus J. Meade [1,†], Kalina M. Manoylov [2] and A. Bruce Cahoon [1,*]

[1] Department of Natural Sciences, The University of Virginia's College at Wise, 1 College Ave., Wise, VA 24293, USA; gcp5a@uvawise.edu (G.C.R.P.); mjm6mv@uvawise.edu (M.J.M.)
[2] Department of Biological and Environmental Sciences, Georgia College and State University, Milledgeville, GA 31061, USA; kalina.manoylov@gcsu.edu
* Correspondence: abc6c@uvawise.edu; Tel.: +1-276-328-0201
† These authors contributed equally to this project.

Citation: Proulex, G.C.R.; Meade, M.J.; Manoylov, K.M.; Cahoon, A.B. Mitochondrial mRNA Processing in the Chlorophyte Alga *Pediastrum duplex* and Streptophyte Alga *Chara vulgaris* Reveals an Evolutionary Branch in Mitochondrial mRNA Processing. *Plants* **2021**, *10*, 576. https://doi.org/10.3390/plants10030576

Academic Editor: Nunzia Scotti

Received: 18 February 2021
Accepted: 13 March 2021
Published: 18 March 2021

Abstract: Mitochondria carry the remnant of an ancestral bacterial chromosome and express those genes with a system separate and distinct from the nucleus. Mitochondrial genes are transcribed as poly-cistronic primary transcripts which are post-transcriptionally processed to create individual translationally competent mRNAs. Algae post-transcriptional processing has only been explored in *Chlamydomonas reinhardtii* (Class: Chlorophyceae) and the mature mRNAs are different than higher plants, having no 5′ UnTranslated Regions (UTRs), much shorter and more variable 3′ UTRs and polycytidylated mature mRNAs. In this study, we analyzed transcript termini using circular RT-PCR and PacBio Iso-Seq to survey the 3′ and 5′ UTRs and termini for two green algae, *Pediastrum duplex* (Class: Chlorophyceae) and *Chara vulgaris* (Class: Charophyceae). This enabled the comparison of processing in the chlorophyte and charophyte clades of green algae to determine if the differences in mitochondrial mRNA processing pre-date the invasion of land by embryophytes. We report that the 5′ mRNA termini and non-template 3′ termini additions in *P. duplex* resemble those of *C. reinhardtii*, suggesting a conservation of mRNA processing among the chlorophyceae. We also report that *C. vulgaris* mRNA UTRs are much longer than chlorophytic examples, lack polycytidylation, and are polyadenylated similar to embryophytes. This demonstrates that some mitochondrial mRNA processing events diverged with the split between chlorophytic and streptophytic algae.

Keywords: mitochondria; RNA processing; algal evolution; circular RNA; polycytidylation; PacBio Iso-Seq

1. Introduction

Mitochondria are membrane-bound organelles known for supplying eukaryotic cells with energy through ATP to carry out cellular functions. This occurs due to aerobic respiration whereby pyruvate is oxidized to CO_2 to generate reduced cofactors that drive the electron transport chain to chemiosmotically fuel ATP synthesis [1]. Despite the crucial role mitochondria have in supplying energy necessary for cellular functions and ATP for other biochemical pathways, it did not originate as a component of the eukaryotic cell. During the late 20th century, the theory of endosymbiosis became widely accepted and states that an aerobic bacterium was absorbed by, and formed an endosymbiotic relationship with, a pre-eukaryotic cell [2,3]. Though it became fully integrated into the Last Eukaryotic Common Ancestor (LECA), the proposed alpha-proteobacterium [4] maintained a portion of its circular genome carrying a conserved set of genes enabling the quick modulation of crucial energy acquisition proteins [5]. This remnant of the bacterial genome is referred to as mtDNA, the mitochondrial genome or chondriome.

Mitochondria have retained their own gene expression machinery, combining bacterial-like traits with novel features that evolved in the host cell [6]. Quite a bit is known about mitochondrial transcription and RNA processing from the compact chondriomes of humans and mice, which can serve for an overview of the process. Briefly, a nuclear-encoded RNA polymerase similar to those found in T3 and T7 bacteriophages [7] recognizes a promoter on both strands of mtDNA with the aid of two transcription factors [8,9]. These promoters occur in the only non-coding region (hyper-variable) and produce two long poly-cistronic primary RNAs known as heavy and light [10–14] with the aid of an elongation factor [14]. Individual mRNAs and tRNAs are removed from the primary transcripts by endonucleolytic cleavage by the enzymes RNaseP and RNaseZ, which precisely remove tRNAs, leaving most of the mRNAs as individual mRNAs with very short 3' and 5' UnTranslated Regions (UTRs), a process called the Punctuation Model [15–17]. Endonucleolytic processing between mRNAs with no intervening tRNA and between an mRNA with an adjacent antisense tRNA has been documented [13], but the enzymatic mechanisms responsible for these processing events are currently unknown. Having no 5' UTRs, these mRNAs lack canonical ribosome binding sequences and use an alternative ribosome binding mechanism that is unique to mitochondria [18]. Once cleaved from the primary transcript, mRNAs may be polyadenylated, which adds the final adenine in some transcripts' stop codons, stabilizes some, and acts as a degradation signal for truncated messages [19–21]. mRNA fragments, but not full-length mRNAs, may also be circularized [22].

The mitochondrial genomes of plants (embryophytes) are much larger than those in animal cells due to expansive intergenic regions, repetitive DNA, and introns [23]. Plant mitochondria share some transcriptional processes with vertebrates. Transcription in plant mitochondria is catalyzed by one or more nuclear encoded phage-like RNA polymerases [24,25], and transcription factors similar to those used by vertebrates are encoded in plant nuclear genomes [26], but their functions have yet to be demonstrated. Due to their sizes, plant chondriomes have multiple promoters dispersed throughout the chondriome [27], yielding multiple primary poly-cistronic transcripts. Post-transcriptional processing takes on an expanded role in plants requiring numerous RNA Processing Factors (RPFs) that target endo- and exo-nucleolytic enzymes, define mRNA ends, and modify transcripts [28]. The 5' termini of genes directly downstream of a transcriptional promoter are formed by the initial nucleotide added by the RNA polymerase [29]. For downstream genes in poly-cistronic transcripts, endonucleolytic cleavage between two genes will simultaneously produce the 5' UTR of one gene and the 3' UTR of an adjacent one. The lengths of these UTRs range from dozens to thousands of nucleotides consistent with the large intergenic regions of plant chondriomes [29–31]. To date, the best-defined cleavage mechanisms in plants are the precise removal of tRNAs by RNaseZ and PRORP. Similarly, tRNA-like secondary structures called t-elements also define intergenic cleavage sites recognized by endonucleases [31–35]. Most protein-coding genes are not separated by tRNAs, and their intergenic cleavage mechanism is hypothetical at this time but involves at least two nucleases [30,36]. Multiple 5' termini for each gene usually result from these processes [30,31]. The 3' ends are less variable and gene specific RPFs bind to them, presumably defining and stabilizing them [37–41]. The prevalence of group I and II introns in plant mitochondria creates an added layer of post-transcriptional processing. Neither class of intron is able to self-splice, so a group of nuclear-encoded RPFs are necessary for their removal [42]. In addition to the major construction of the mRNA coding regions, individual nucleotides are modified in a process known as RNA editing, which is common in higher plant mitochondria and chloroplasts [43]. Once the mRNA is no longer needed, it may be marked for degradation by way of polyadenylation by nuclear encoded factors in a manner similar to that of bacteria [44–47].

Our understanding of mitochondrial transcription and RNA processing in algal species is mostly limited to the single-celled photosynthetic green alga *Chlamydomonas reinhardtii* P.A. Dangeard, which is a well-established model system [48]. *C. reinhardtii* has a small linear chondriome [49] that is unusual among algae but is a conserved trait among the Reinhardtinia clade of the Order Chlamydomonadales [50,51]. In this species, transcription is initiated on each of the two strands from promoters in a small intergenic region to produce two primary transcripts [52,53]. Each mRNA is endonucleolytically cleaved directly adjacent to the AUG start codon, leaving no 5′ UTR, similar to those seen in animal systems. The 3′ UTRs are comprised of various lengths of template-derived intergenic regions and may have non-template polycytosine and/or polyuracil tails added, presumably as part of the maturation process [54–56]. The poly-cytidylation of mitochondrial mRNAs seen in green algae is unusual and appears to be limited to the algal class Chlorophyceae [55]. It has been hypothesized that these leaderless mRNAs use an alternative ribosome-binding mechanism, but there is evidence that the mature mRNAs are circularized, which brings putative ribosome binding sites (RBSs) located in the intergenic regions of the *Chlamydomonas* chondriome upstream of the start codon to initiate translation [56]. mRNAs in *C. reinhardtii* are also poly-adenylated, which serves as a degradation signal consistent with mitochondria in other eukaryotes and bacteria [54,57,58]. mRNA editing, which is common among embryophytes, is missing in both the chlorophyte and streptophyte lineages of green algae [59], suggesting that some post-transcriptional processes were acquired by embryophytes after they invaded land.

The purpose of this study was to define the 5′ and 3′ UTRs of mitochondrial mRNAs in two algae, *Pediastrum duplex* Meyen (P: Chlorophyta, C: Chlorophyceae, F: Hydrodictyaceae) from the chlorophyte algal clade and *Chara vulgaris* Linnaeus (P: Charophyta, C: Charophyceae, F: Characeae) of the charophyte algal clade. *P. duplex* is a member of the same class as *C. reinhardtii*, but it has a circular chondriome that is several times larger [60,61], making the architecture more similar to that found in other algae. By defining the termini, we hoped to determine if RNA processing events seen in *C. reinhardtii* are also used among other chlorophytic algae or are related to its compact genome. *C. vulgaris* has a mitochondrial genome similar in size to *P. duplex*, but the gene content and synteny are more similar to bryophytes [62,63]. We analyzed the 3′ and 5′ ends from *C. vulgaris* to see if mRNA end processing resembled higher plants or chlorophytic green algae. Circular RT-PCR (cRT-PCR) and PacBio long-read sequencing were used to define the transcript termini of 12 mitochondrial mRNAs from each species. cRT-PCR allows the mapping of both the 3′ and 5′ ends of an RNA by artificially ligating them together followed by the production of a cDNA across the ligation site, PCR amplification of the sequences flanking the ligation site and sequencing of those amplicons. PacBio Iso-Seq is a long-read RNA sequencing technology which can sequence full-length RNAs including their 3′ and 5′ termini. This platform sequences poly-adenylated RNAs so organellar mRNAs must be artificially poly-adenylated to increase the likelihood they will be sequenced.

2. Results

2.1. Pediastrum duplex

In *P. duplex*, *cox2a*, *cox3*, *nad2*, *nad4*, and *nad4L* 5′ termini occurred directly upstream of the AUG start codon, essentially leaving no UTR (Table 1). For *atp6*, 5′ and 3′ end processing occurred within a 9 nt genomic DNA (gDNA) encoded stretch of adenines flanking the gene (Figure 1A). Since the same oligonucleotide sequence occurred at both ends, it was not possible to distinguish the exact location of the 5′ or 3′ exonucleolytic cleavage using cRT-PCR. Two of the *P. duplex* mRNAs (*atp9* and *nad1*) had cleavage sites producing 5′ UTR termini downstream of the predicted start codons in archived chondriome maps (KR026340, KR026340, MK895949). For *atp9* (Figure 1B) and *nad1* (Figure 1D), there is an in-frame AUG start codon adjacent to the cut site leaving a short 5′ UTR consistent with the other genes. For *cob*, there was disagreement in the 5′ terminus between the circRT-PCR and PacBio techniques. Using circ-RT-PCR, a single 5′ terminus 22 nt downstream of the predicted start codon was detected (Figure 1C), while reads using PacBio IsoSeq revealed a single terminus adjacent to the originally predicted start codon.

Table 1. Site of 5′ UTR terminus upstream from start codon (in nucleotides) in *Pediastrum duplex* mitochondria, nd = no data. * Some 5′ UTR termini detected in this study occur downstream of the start codons in archived chondriomes. The distances presented in this table are marked from the next available AUG start codon.

	Site of 5′ UTR Terminus Upstream from Start Codon (in Nucleotides)	
Gene	Circular RT-PCR	PacBio
atp6	−2–11	nd
atp9	−3 *	nd
cob	−2 *	0
cox1	nd	0
cox2a	0	nd
cox3	0	0
nad1	−15 *	nd
nad2	0	nd
nad4	0	0
nad4L	0	0
	−81–93	nd
nad5	nd	nd
nad6	nd	0

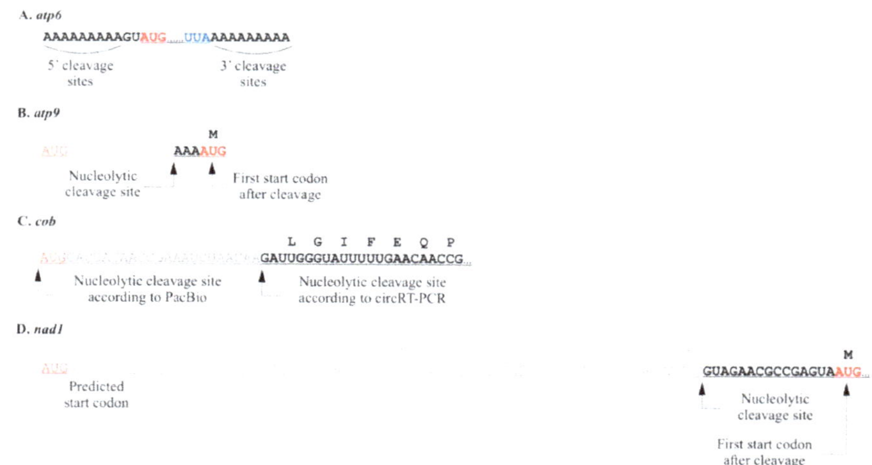

Figure 1. The 5′ termini and UTRs of four *P. duplex* mitochondrial mRNAs. Start codons are red, stop codons blue, and truncated portions of coding regions grey. (**A**). The 5′ UTR of the *atp6* mRNA occurred within a 9 nt templated stretch of adenines that also appears downstream of the stop codon. (**B**). The *atp9* 5′ terminus occurred downstream of the AUG start codon in archived *P. duplex* chondriomes, suggesting the protein may be six amino acids shorter than predicted. (**C**). Two *cob* 5′ termini were detected. Circular RT-PCR revealed one downstream of the AUG start codon in archived *P. duplex* chondriomes, while the PacBio Iso-Seq technique revealed a terminus directly adjacent to the predicted start codon. (**D**). The *nad1* 5′ terminus occurred downstream of the AUG start codon in archived *P. duplex* chondriomes, suggesting the protein may be thirty amino acids shorter than predicted.

P. duplex 3′ UTR lengths were gene specific, with several having two or more termini (Figure 2A–J and Table 2). Most were relatively short, fewer than 25 nts, the exception being *cob* which were 100 and 110 nts in length (Figure 2C). Eight genes (*atp6*, *atp9*, *cob*, *cox1*, *cox2a*, *cox3*, *nad2*, and *nad5*) were polycytidylated. For some, this occurred at specific termini, e.g., *atp6*, *atp9*, *cob*, *cox1*, *cox2a*, and *cox3*. For two genes, *nad2* and *nad5*, the poly(C) additions occurred at variable locations within templated repetitive AU regions beginning 9nts downstream of the stop codons (Figure 2G,J and Table 2). No poly(C) additions were detected on *nad1*, *nad4*, or *nad4L*. On *nad4L*, there was an AU repeat region adjacent to the stop codon, but no poly(C) additions were detected. There was general agreement of the 3′ termini of fully processed mRNAs between cRT-PCR and PacBio reads for all but one transcript, *nad4L*. For this gene, cRT-PCR provided one 3′ terminus with a truncated stop codon (Figure 2J upper sequence), whereas PacBio data provided a 3′ terminus 37 nt downstream (Figure 2J lower sequence). The nucleotide sequences between the stop codon and the 3′ terminus for each gene were aligned and analyzed using a logo plot, and the 15 nucleotides upstream of the terminus were comprised nearly exclusively of adenines and uracils (Supplemental Figure S1A). These regions were analyzed for RNA secondary structure and none were detected.

Naturally circularized mRNAs were also detected using RT-PCR. In *P. duplex*, circularized variants carrying full-length coding regions were detected for seven mRNAs (*atp6*, *atp9*, *cob*, *cox2a*, *cox3*, *nad4L*, and *nad5*) (Figure 3). For five of these, the circularization coincided with a tandemly repeated nucleotide motif. The ligation site for naturally circularized *atp6* transcripts occurred within a stretch of template coded adenines (Figure 3A). Several circularized variants of *cob* were found (Figure 3C), one where the circularization occurred within a template encoded polyU motif and a second within two AU rich motifs. Two *cox3* circular transcript variants were detected, one where the ligation occurred within a GAACGAA motif and a second ligated at a GCGTCTT motif that removed the final 45 nts of the coding region. Two *nad4L* circular variants were detected, each occurring within AT rich motifs and including full-length coding regions. Three naturally circularized transcripts were detected with poly(C) additions. The two circular variants of *atp6* and *cob* with a poly(C) addition had severely truncated coding regions (Figure 3A,C). Two *atp9* circular variants were detected and both had poly(C) additions and full-length coding regions (Figure 3B). A single circularized variant of the *cox2a* transcript was detected, with no obvious repeat motif and no poly(C) additions (Figure 3D).

The long-read PacBio data did not cover the entire chondriome or contain reads longer than ~2200 nt but, when combined with the circ-RT-PCR data, they did allow the detection of some broader transcript processing events from three portions of the *P. duplex* chondriome (Figure 4). For *cox1*, reads spanning the two exons and the intron were detected (Figure 4A). All of these reads had 5′ termini directly adjacent to the start AUG, while some were polycytidylated on the 3′ terminus. Transcripts with the intron removed were also detected that had the same 5′ and 3′ termini as the unspliced transcript. PacBio reads were also produced for another section with three genes, *nad2-nad6-cob*, flanked by tRNAs (Figure 4B). For *nad2*, transcripts appeared to have been endonucleolytically cleaved adjacent to *trnN*, forming a 5′ terminus for *nad2* and adjacent to *nad6* to form a 3′ terminus. The 5′ end was further processed, leaving multiple termini eventually resulting in an mRNA with no UTR that was also polycytidylated on the 3′ terminus. A transcript with both *nad6* and *cob* was detected. Its 5′ terminus occurred adjacent to *nad6*'s start codon, while the 3′ terminus appeared to have been created by the cleavage of *trnV* from the primary transcript followed by polycytidylation. The *cob* coding region was cleaved away from *nad6*, leaving two different 5′ termini. The linear transcripts were polycytidylated, whereas the circular version of *cob* was ligated with no poly(C) tract. No individual *nad6* transcript was detected from either circRT-PCR or PacBio results. A third region with the two genes *nad4L-atp9* flanked by tRNAs was also produced by the PacBio Iso-Seq methodology (Figure 4C). A single transcript that appeared to have been produced by the removal of the two tRNAs from the primary transcript was detected. The 3′ terminus of this poly-cistronic mRNA was poly-citidylated after *trnE* was removed. The removal of the *trnG* occurring 5′ of *nad4L* left 170 nt upstream of the start codon, but this was removed, leaving a 5′ terminus adjacent to the start AUG. The 3′ terminus was produced by endonucleolytic cleavage, leaving a 37 nt 3′ UTR that was polycytidylated. A second *nad4L* transcript was detected with a longer 5′ UTR (−80–94 nt within an AU repeat region) and a 3′ terminus comprised of a truncated stop codon that had been polycytidylated. Both versions of these shortest *nad4L* transcripts were detected as circular RNAs and neither contained a poly(C) tract. The *atp9* coding region was cleaved from the primary transcript, leaving a 5′ terminus −2 nt upstream of its start AUG, but its 3′ terminus was the one formed by the removal of *trnE*. A non-polycytidylated version of this mRNA was circularized.

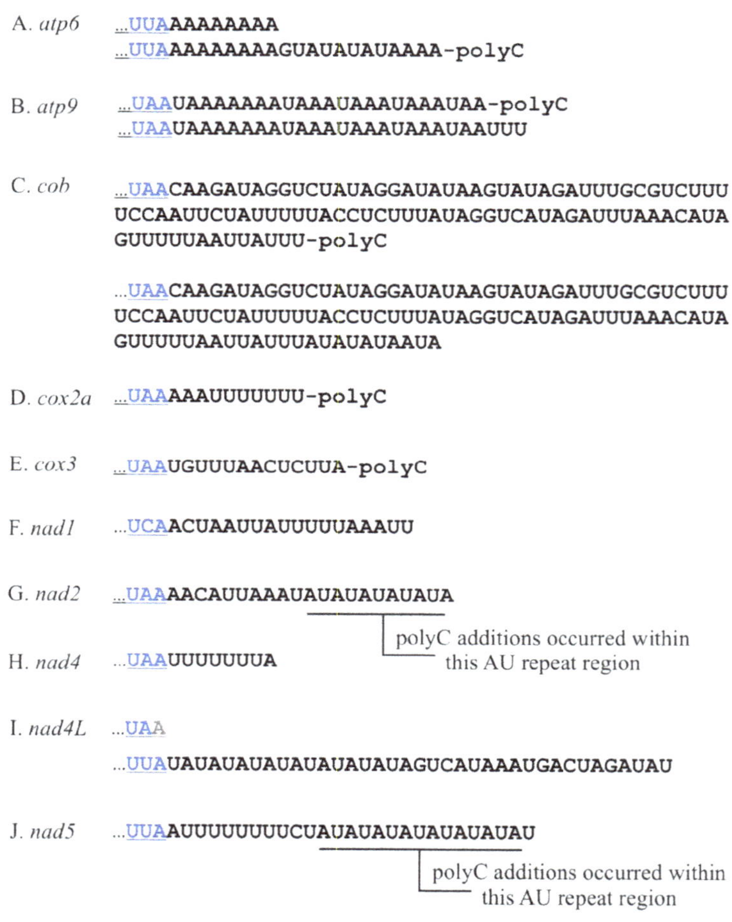

Figure 2. The 3′ termini and UTRs of ten *P. duplex* mitochondrial mRNAs. Stop codons are blue. (**A**). Two 3′ termini were detected for *atp6*. The upper (shorter) sequence had no oligonucleotide addition, while a portion of transcripts represented by the lower (longer) sequence had poly(C) additions. (**B**). For *atp9* a portion of the transcripts with the upper terminus had a poly(C) addition, while the slightly longer one represented by the lower sequence did not. (**C**). For *cob* a portion of the transcripts with the upper terminus had a poly(C) addition, while the one represented by the longer sequence did not. (**D**). For *cox2a* a single terminus was detected and a portion of them had a poly(C) tail. (**E**). A single terminus, some of which had a poly(C) addition, was detected for *cox3*. (**F**). A single terminus was detected for *nad1* and no oligonucleotide additions were detected. (**G**). For *nad2* poly(C) additions were detected at several different termini within an AU repeat region. (**H**). A single terminus was detected for *nad4* and no oligonucleotide additions were detected. (**I**). A single terminus was detected for *nad4L* and no oligonucleotide additions were detected. (**J**). For *nad5* poly(C) additions were detected at several different termini within an AU repeat region.

Table 2. Lengths of the most abundant 3′ UTRs found in *Pediastrum duplex* and those found with a non-template addition

| | Circular RT-PCR | | | | PacBio Long Reads | | |
Gene	Length of 3′ UTR from Stop Codon	UTRs with polyC Addition	UTRs with polyA + polyC Addition	UTR with polyA Addition	Length of 3′ UTR from Stop Codon	UTRs with polyC Addition	Genes with an Adjacent tRNA (Distance from Stop Codon in nt)
atp6	21 0–9	+ -	- -	- -	nd nd	nd nd	
atp9	22 25	+ -	- -	- -	16–22 nd	+ nd	26
cob	110 100	- +	- -	- -	104–110 nd	+ nd	110
cox1	13	+	-	-	7–14	+	
cox2a	10	+	-	-	nd	nd	28
cox3	13	+	-	-	10–13	-	
nad1	18	-	-	-	nd	nd	
nad2	9–21	+	-	-	5–23	-	
nad4	8	-	-	-	7–8	+	786
nad4L	−1	-	-	-	37	-	
nad5	0–8 9–24	- +	- +	+ 	1–3	-	42
nad6	nd	nd	nd	nd	nd	nd	

nd: no data.

Since tRNA removal was found to be integral in the maturation of *cob* and *atp9* 3′ termini, the placement of tRNAs was compared to the 3′ termini of other genes (Table 2). Five of the genes we analyzed had a 3′ adjacent tRNA, but only two, *cob* and *atp9*, had mature 3′ ends matching the placement of those tRNAs. The possibility of t-elements was considered for the other genes, but no evidence of secondary structures immediately downstream of the mature 3′ ends was detected for *atp6, cox1, cox2a, cox3, nad1, nad2, nad4, nad4L, nad5,* or *nad6*.

A. *atp6* AAAAAAAGUAUG...UAAAAAAAAA

AUG

U...UAAAAAAAAAGUAUAUAUAAAA-polyC

B. *atp9*

M
AAAAAUG...UAAUAAAAAAAUAAAUAAAUAA-polyC

GCAAAAAUG...UAAUAAAAAAAUAAAUAAAUAAAUAA-polyC

C. *cob* GUUUUUUUGUUUUUUAUG...UAACAAGAUAGGUCUAUAGGAUAUAAGUAUAGAUUUGCGUCUUUUCCAAUUCUAUUUUUUACCUCUUUAUAGGUCAUA
GAUUUAAACAUAGUUUUUU

AUUAUUUUUAUAAAUAAGUUUUUUUUGUUUUUUUAUG...UAACAAGAUAGGUCUAUAGGAUAUAAGUAUAGAUUUGCGUCUUUUCCAAUUCUAUUUUUUA
CCUCUUUAUAGGUCAUAGAUUUAAACAUAGUUUUUAAUUAUUUAUAUAUAAUA

AUG
CAAGAUAGGUCUAUAGGAUAUAAGUAUAGAUUUGCGUCUUUUCCAAUUCUAUUUUUUACCUCUUUAUAGGUCAUAGAUUUAAACAUAGUUUUUAAUU
AUUUAUAUAUAAUA

       └── polyA additions occurred
             within this region

D. *cox2a* CGAAGAAGAAGCCGAUGCGCAUCGGCUUCUUCGCUUCGCGAACGAAGAAGCCAGCCGAAGCGAAAGAAAGCAGGGUGCGGGGGUGCACUUUGCUGU
CGCGGAGCGAAUGCAAACAAGAAAUAUUGAGCUUCUUAUAAAGCUGGGUAUUUAUUAUUAUG...UAAAAAUUUUUUUAUAUAUUAAAUUAUUUGAGU
GUGUAGCUCAAUUGGUUAGAGCGCAGGACCGAUAAUCCUGAGGUUGCAGGUUCAAUCCCUGCCAUACUCAAACAAUGGCGUUAGCACUGGCUUCUU
UCGCUUCGGCUGGCUUCUUCGUUCGCUUGCGCUUUCG

AGCCGAUGCGCAUCGGCUUCUUCGCUUCGCGAACGAAGAAGCCAGCCGAAGCGAAAGAAAGCAGGGUGCGGGGGUGCACUUUGCUGUCGCGGAGCG
AAUGCAAACAAGAAAUAUUGAGCUUCUUAUAAAGCUGGGUAUUUAUUAUUAUG...UAAAAAUUUUUUUAU

E. *cox3* GAACGAAGAAGCACCAGUUCUACGAAUCGCGUCUUGUCAGACGCGGGCAAGUCCUACGGACCUUUUAAAUUUUUUUUUUUGUUUUUUAUG...UAAUGU
UUAACUCUUAUUUUUAUUUAUAAAAGUAGAGUAGUAUAAAAGUGGAGCCUAUAUAUAUAUAAUAUAUAAGUCAAAAACUAGAGUACUAGUAGUACU
CUAACUUUUCCACUUUGCUGUGUAUCUAAAAUUUUGAAAUGGGUACAGUGAGCUUAUUUAUUUUUUUUUUUUAUAUAUUUAUUUUUUGUUUUUAUUGACG
UUAUAUUUGGAGCACCCGAAAGCAGGUGAUCUUCCCUUGUCCAAGUUGAAGGCUGGCUUUUUGCACCGCUGGAGGACUGAACUCGUGUACGUGGUA
AAGUACUGAGGUGAGGUGGGGGAAGGAGUGAAAGGCUAAUCAAACUUGCCUAUCGCUGGUUUUUGAACGAA

GCGUCUUGUCAGACGCGGGCAAGUCCUACGGACCUUUUAAAUUUUUUUUUUUGUUUUUUAUG...GCGUCUU
UAA

F. *nad4L* UAUAUAUAUAUAUACAAAAUUUACAACUAAAUUUAUCUUUAUAUUAAUUAAAAGAUAAAUGUGUUUAUCUGUACUUUUAUUUAUAAAAGUACAGAU
G...UAAUAUAUAUAUAUAUAUAUA

UUUAUAAAAGUACAGAUG...UAAUAUAUAUAUAUAUAUAUAGUCAUAAAUGACUAGAUAUAAAACUGCAUUUAUAAA

G. *nad5* UAUAUAUG...UAAUUUUUUUUUCUAUAUAUAUAUAUAUAUA

Figure 3. Naturally circularized mRNAs found in *P. duplex* mitochondria. Start codons are colored red, stop codons blue, repeat sequences orange, and truncated portions of coding regions grey. (**A**). Two circularized mRNAs were detected for the gene *atp6*, the upper sequence represents a full-length coding region whose 3′ and 5′ termini were ligated within templated adenine stretches that flank the coding region. The lower sequence represents a circularized portion of the mRNA. (**B**). Two versions of a circularized *atp9* transcript were detected. Both would be full length considering a start codon downstream of the previously predicted one (grey) would be the actual start codon. (**C**). Two full-length circularized versions of the *cob* transcript were detected (upper two sequences). The ligation termini coincided with two different repeat sequences (orange). A third circularized transcript with a truncated coding region was also detected for *cob*. (**D**). Two circularized full-length coding regions were detected for *cox2a*. (**E**). Two circularized *cox3* transcripts were detected, both ligations having occurred at repeat sequences. One carried a full-length coding region (upper) the other truncated (lower). (**F**). Two full-length circularized versions of the *nad4L* transcript were detected where ligation occurred at AU-rich repeat sequences. (**G**). One circularized *nad5* transcript was detected with the ligation occurring within AU-rich repeat sequences.

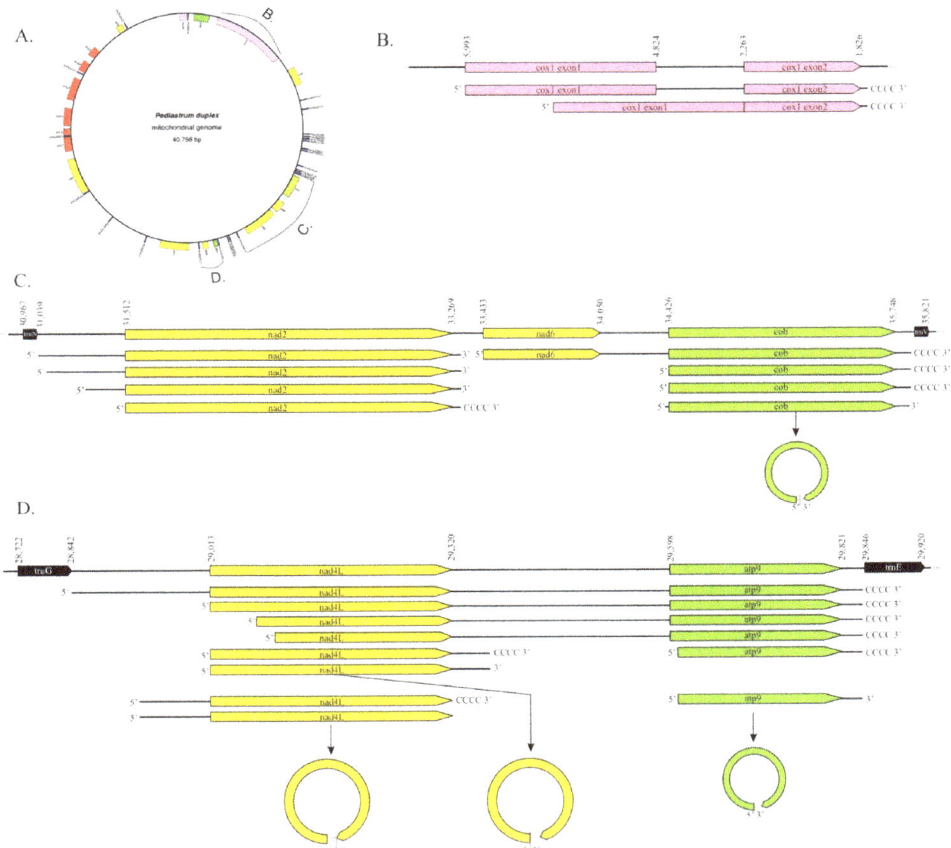

Figure 4. Mitochondrial RNA processing as determined by PacBio Iso-Seq and circRT-PCR. (**A**). The mitochondrial genome of *P. duplex* (GenBank MK895949). The regions highlighted in portions B, C, and D of this figure are marked. The map was generated using OGDRAW [64]. (**B**). The *cox1* gene, the upper diagram represents a portion of the chondriome, the lower two portions represent partially and fully processed transcripts. (**C**). The *trnG-nad2-nad6-cob-trnE* portion of the chondriome. (**D**). The *trnG-nad4L-atp9-trnE* portion of the chondriome.

2.2. Chara vulgaris

The 5′ UTRs were much longer than those observed in *P. duplex* (Table 3). Based on the termini we detected, UTRs ranged from 6–273 nucleotides, with an average length of 80 nts (S.E. = 15 nt). They were also more variable, with two or more termini detected for seven genes, *atp6*, *cob*, *cox2*, *cox3*, *nad1*, *nad4*, and *nad4L*. For three of the genes, *cox1*, *nad1*, and *nad2*, the mapped 5′ termini occurred downstream of the start codons in Gen-Bank record NC_005255 (Figure 5). For *cox1*, a single terminus was detected 125–129 nts downstream of the predicted start codon. The next start AUG occurs 75–79 nt downstream (Figure 5A). Three 5′ UTRs were detected for *nad1*, all of which remove the predicted start codon but, depending upon the cleavage site, leave two possible alternative AUG start codons (Figure 5B). In *nad2*, two 5′ UTRs were detected, both of which leave a single alternative AUG start codon (Figure 5C). The length of the 5′ UTRs in *C. vulgaris* raised the possibility that they may fold to form RNA secondary structures, but the probability of secondary structures in the 5′ UTRs was found to be extremely low. The coverage of chondriome derived transcripts for *C. vulgaris* using PacBio sequencing was very low, so the 5′ termini of only two genes was recovered, *atp9* and *cox2*. Neither were the same length as those detected by circRT-PCR (Table 3).

Table 3. Site of 5′ UTR terminus upstream from start codon (in nucleotides) in *Chara vulgaris* mitochondria, nd = no data. * 5′ UTR termini detected in this study occur downstream of the start codons found in archived chondriomes. The distances presented in this table are marked from the next available AUG start codon.

	Site of 5′ UTR Terminus Upstream from Start Codon (in Nucleotides)	
Gene	**Circular RT-PCR**	**PacBio**
atp6	−76 −62	nd
atp9	−56	−93
cob	−161 −46	nd
cox1	−150 *	nd
cox2	−318 −18	−52
cox3	−153 −31	nd
nad1	−73 * −27 * −8 *	nd
nad2	−112 *	nd
nad4	−273 −107 −99 −82 −18 −6	nd
nad4L	−260 −18	nd
nad5	nd	nd
nad6	−36	nd

C. *vulgaris* 3′ ends ranged from 0–162 nts with an average of 61.3, S.E. = 7.5, (Table 4, Figure 6). Multiple 3′ termini were detected for each gene, except *cox1*, and *nad2*, which had single termini. Polyadenylation was detected on eight genes (*atp6*, *atp9*, *cob*, *cox3*, *nad1*, *nad2*, *nad4*, and *nad4L*). For genes where multiple termini were detected, the polyA tail only occurred on one of those termini (Figure 6B,C,F,G,I). The exception was *atp6*, where all three termini were polyadenylated (Figure 6A). The proportion of those specific transcripts with polyA additions varied considerably. For example, 0.7% of *atp9* transcripts with 52–57 nt 3′ UTRs had a polyA tail, whereas the majority of specific *nad2*, *nad4*, and *nad4L* transcripts were tailed. PacBio sequencing produced data for four genes and the 3′ termini agreed with three (Table 4). The exception was *cob*, where the PacBio sequencing revealed a longer 3′ UTR than those found with circRT-PCR. PacBio sequencing could not be used to detect non-template poly(A) tails since the mitochondrial transcripts were artificially polyadenylated to accommodate the Iso-Seq technique. The forty nucleotides upstream of the 3′ termini were analyzed for conserved sequences using a logo plot and none were found (Supplemental Figure S1B). The length of the 3′ UTRs in C. *vulgaris* raised the possibility that these regions could fold into secondary structures. Secondary structure prediction suggested a high probability that stable stem-loop structures occur adjacent to the 3′ terminus in all but one (*cox2*) of these 3′ UTRs (Supplemental Figure S2).

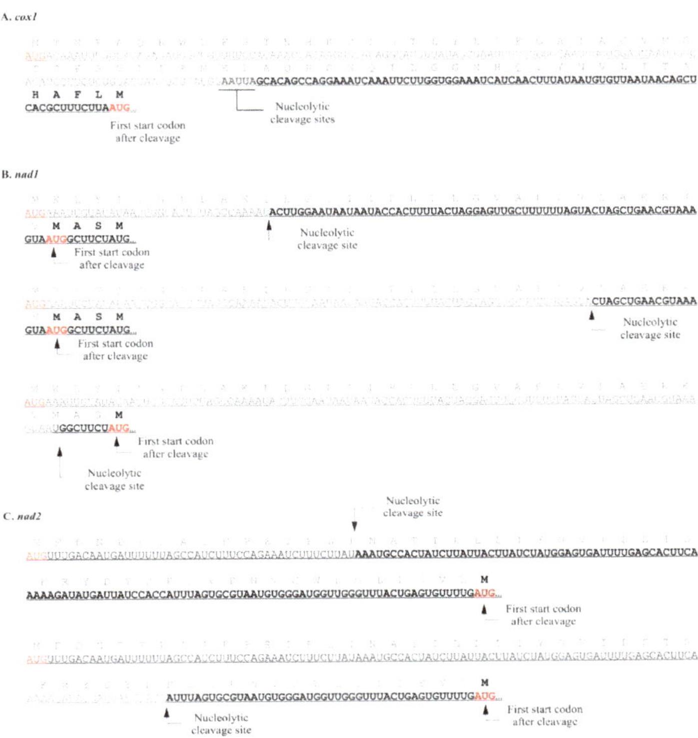

Figure 5. The 5′ termini and UTRs of three *C. vulgaris* mitochondrial mRNAs where the 5′ terminus occurred downstream of predicted start codons. Start codons are red and truncated portions of coding regions grey. (**A**). For *cox1* several 5′ termini were detected within a 5-nucleotide region. (**B**). Three 5′ termini were detected for *nad1*. (**C**). Two different termini were detected for *nad2*.

Table 4. Length of the most abundant 3′ UTRs found in *Chara vulgaris* and those with a non-template polynucleotide addition.

	Circular RT-PCR		PacBio		
Gene	Length of 3′ UTR from Stop Codon	UTRs with polyA Addition	Length of 3′ UTR from Stop Codon	Genes with an Adjacent tRNA (Distance from Stop Codon in nt)	Evidence of RNA Secondary Structure Immediately Downstream of 3′ Terminus
atp6	79 81 83	+ + +	nd	459	+
atp9	52–57 58	+ -	56	92	-
cob	54 68	- +	121		-
cox1	73	-	73		
cox2	51 82 107 162	- - - -	nd		-
cox3	23 51 56	+ - -	52	6056	-

Table 4. *Cont.*

	Circular RT-PCR		PacBio		
Gene	Length of 3′ UTR from Stop Codon	UTRs with polyA Addition	Length of 3′ UTR from Stop Codon	Genes with an Adjacent tRNA (Distance from Stop Codon in nt)	Evidence of RNA Secondary Structure Immediately Downstream of 3′ Terminus
nad1	20	-	nd		+
	67	-			
	70	+			
	141	-			
nad2	87	+	nd	1189	+
nad4	10	+	nd		-
	13	-			
	21	-			
nad4L	0	-	nd	2508	-
	29	+			
nad5	nd	nd	nd		
nad6	nd	nd	nd	5	

nd: no data.

A. *atp6*   …UAACCACUUUCACUUCGUGGCCUCUGGUGACACAACCGCCGCGAAGCGUUCAUCUUUCCCACCCACCCCAGGGUGGGGUGGU-polyA
…UAACCACUUUCACUUCGUGGCCUCUGGUGACACAACCGCCGCGAAGCGUUCAUCUUUCCCACCCACCCCAGGGUGGGGUGGUGU-polyA
…UAACCACUUUCACUUCGUGGCCUCUGGUGACACAACCGCCGCGAAGCGUUCAUCUUUCCCACCCACCCCAGGGUGGGGUGGUGUCG-polyA

B. *atp9*   …UAAUGAAAUUAUGUACCCGUAAAAAAAUGGCCGGGAAAUCCUGGCCUUACGGGUACCACCCU
…UAAUGAAAUUAUGUACCCGUAAAAAAAUGGCCGGGAAAUCCUGGCCUUACGGGUACCACCC

polyA additions occurred within this region

C. *cob*   …UAGCUUUCAUUUUGAAUUGAAUGUCAUGCAUGAUAUGUACCAAUUACCCACCCGUUCGGG
…UAGCUUUCAUUUUGAAUUGAAUGUCAUGCAUGAUAUGUACCAAUUACCCACCCGUUCGGGUGGGGUUUCAA-polyA

D. *cox1*   …UAAUUGUUGUUAUAAGAAAGUAGUCCGUUUUUAUUAUAUUACAUUGCAUUGCCCCCAUUUAACUUGGGGGCAACGA

cox3

E. *cox2*   …UAAAAAACAAAGGAGCGAAGCAAUGAGUAUGAGUUCUCAAAAGCAUCCUUAUCA
…UAAAAAACAAAGGAGCGAAGCAAUGAGUAUGAGUUCUCAAAAGCAUCCUUAUCAUUUAGUUGAUCCAAGUCCAUGGCCUAUUUUU
…UAAAAAACAAAGGAGCGAAGCAAUGAGUAUGAGUUCUCAAAAGCAUCCUUAUCAUUUAGUUGAUCCAAGUCCAUGGCCUAUUUUUGGUUCACUGGGAGCUUUGGCAAGUA
…UAAAAAACAAAGGAGCGAAGCAAUGAGUAUGAGUUCUCAAAAGCAUCCUUAUCAUUUAGUUGAUCCAAGUCCAUGGCCUAUUUUUGGUUCACUGGGAGCUUUGGCAAGUACCAUUGGUGGUGUUAU

F. *cox3*   …UAACACACACUUACCCCGCCCCGGGG-polyA
…UAACACACACUUACCCCGCCCCGGGGCCCUUUACUUUAGGGCGCCGGGUCCACA
…UAACACACACUUACCCCGCCCCGGGGCCCUUUACUUUAGGGCGCCGGGUCCACAGAACA

G. *nad1*   …UAACUUCACUUUGUUAAGUAGUG
…UAACUUCACUUUGUUAAGUAGUGAUUCAUAUAGAAUAGAAAGGUGCCCACUAUAGACGUGGGUGCCACUU
…UAACUUCACUUUGUUAAGUAGUGAUUCAUAUAGAAUAGAAAGGUGCCCACUAUAGACGUGGGUGCCACUUCGC-polyA
…UAACUUCACUUUGUUAAGUAGUGAUUCAUAUAGAAUAGAAAGGUGCCCACUAUAGACGUGGGUGCCACUUCGCCAGAUGUAUUGUAUCUACUUCUACUAGCUACGCUUAACAAAGAAAGAAGUUCCCACUUCCUUCCUUCCUUC

H. *nad2*   …UAAGUUUCGAGCGGAGACGAAACUUCGUAGUAAUAAAAUAAAUAUAAAUACGCGCAGCUAACUAAAUGCCAUAAAGGCUGCGCGUACUCG-polyA

I. *nad4*   …UAAAGGGCCUUUG-polyA
…UAAAGGGCCUUUGUAU
…UAAAGGGCCUUUGUAUUUGACCGG

J. *nad4L*   …UAA
…UAAGAAAGGCGCCUAAAAAAGAGGCGCCUUAG-polyA

Figure 6. The 3′ termini and UTRs of ten *C. vulgaris* mitochondrial mRNAs. Stop codons are blue. (**A**). Three termini were detected for *atp6*, a portion of each had a poly(A) addition. (**B**). Several termini were detected for *atp9*. The longer one (upper) had no oligonucleotide additions, while those lacking the terminal uracil had poly(A) additions within a 6 nt region. (**C**). Two *cob* termini were detected and the shorter one had no detectable oligonucleotide additions while the longer (lower) one did. (**D**). A single terminus was detected for *cob* with no oligonucleotide additions. (**E**). Four termini were detected for *cox2*. All had a portion of the adjacent *cox3* gene and no oligonucleotide additions. (**F**). Three *cox3* termini were detected. A portion of the shortest UTR had poly(A) tails, while the two longer ones did not. (**G**). Four termini were detected for *nad1* but only one had a poly(A) tail. (**H**). A single terminus was detected for *nad2* and a portion of them had a poly(A) tail. (**I**). Three termini were detected for *nad4* and a portion of the shortest had a poly(A) tail. (**J**). Two termini were detected for *nad4L*. One occurred directly adjacent to the stop codon. A portion of the longer UTR had a poly(A) addition.

Since tRNA placement was found to be important for 3′ maturation of some *P. duplex* genes, the distances of mature 3′ termini from the stop codons of genes with an adjacent tRNA were compared for *C. vulgaris* (Table 4). For the twelve genes used in this study, none had 3′ termini formed from the removal of an adjacent tRNA. The presence of RNA secondary structure (t-elements) immediately downstream of 3′ termini was also checked, and three genes (*atp6*, *nad1*, and *nad2*) had potential stem-loop structures adjacent to their mature 3′ termini.

In *C. vulgaris*, circularized full length coding regions were detected for five genes. Circularized *nad2* transcripts matched the 3′ and 5′ termini detected in earlier experiments (Figures 5C and 6H), suggesting that only naturally circularized transcripts were detected for this gene. For the genes *atp9*, *cox1*, *cox2*, and *nad4L*, circularized variants differed from the artificially circularized transcripts analyzed in previous experiments and are represented in Figure 7. For genes *cox2*, *cox3*, *nad1*, and *nad4*, only fragments of coding regions were found circularized. In *C. vulgaris*, there were no repeat motifs associated with the ligation sites and no polyA additions detected in the circularized transcripts.

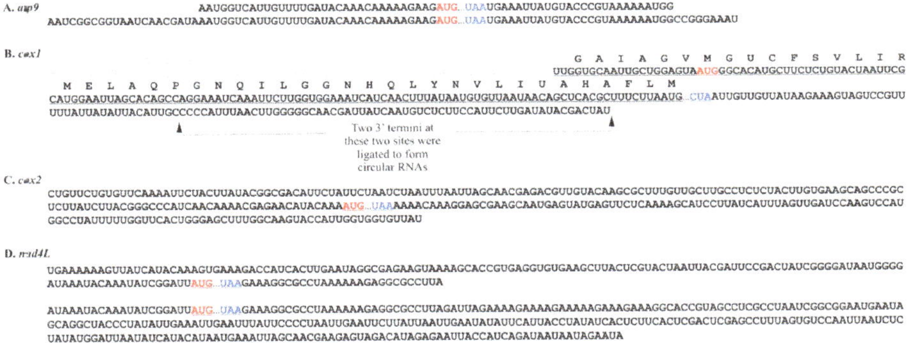

Figure 7. Naturally circularized mRNAs found in *C. vulgaris* mitochondria. Start codons are colored red, stop codons blue, and truncated portions of coding regions grey. (**A**). Two circularized versions of full-length transcripts were detected for *atp9*. (**B**). Two circularized *cox1* transcripts were detected. Both had the same 5′ end but different termini on the 3′ ends. (**C**). A single circularized version of the *cox2a* transcript was detected. (**D**). Two circularized full-length *nad4L* transcripts of different lengths were detected.

3. Discussion

Green algae form two discrete clades, chlorophytes and charophytes [65]. Chlorophytes are morphologically diverse, cosmopolitan and contain the majority of green algae. Chlorophyte chondriomes are nearly as diverse as their morphologies, ranging from 15,500–66,000 bp with few introns and highly variable intergenic spaces, gene content, and gene synteny [60,66–75]. The knowledge of transcription and post-transcriptional processing events in chlorophytic mitochondria is limited to the model system *C. reinhardtii*, where mature mRNAs have no 5′ UTR and relatively short 3′ termini that may be poly-cytidylated. *C. reinhardtii* has a very small linear chondriome that is not representative of the majority of algae, and it is unknown if the post-transcriptional processing characteristics are hallmarks of its reduced chondriome or traits conserved among chlorophytes. Charophytes diverged from chlorophytes a billion or more years ago and have biochemical and morphological synapomorphies shared only with embryophytes [76]. The relatively few extant species belong to six monophyletic clades and are the closest living algal relatives of land plants [63,77]. The chondriomes of charophytes range widely from 56,500 bp to >201,000 bp, with highly variable gene order, density, and intron placement [63].

3.1. Mitochondrial Processing in P. duplex Resembles That Seen in C. reinhardtii

We mapped the 5′ and 3′ termini of presumably mature *P. duplex* mitochondrial mRNAs and found the 5′ UTRs to be very short or non-existent, the 3′ UTRs of varying lengths, the polycytidylation of 3′ termini, and circularized full-length mRNAs. In the only other chlorophyte where mitochondrial RNA processing has been documented, *C. reinhardtii*, individual mRNAs are endonucleolytically cleaved from primary transcripts directly adjacent to the start codons, leaving no 5′ UTR. The remaining intergenic region becomes the 3′ UTR of an adjacent gene and is of varying lengths, presumably due to exonucleolytic processing [56]. It has also been demonstrated that non-template polycytidylation occurs on the 3′ termini of *C. reinhardtii*, and that this phenomenon may be limited to the Phylum Chlorophyta since it has been found in representative species from Peridinophyceae, Prasinophyceae, Trebouxiophyceae, and Chlorophyceae, but not in a red alga (*Chondrus crispus* Stackhouse), a glaucophyte (*Cyanophora paradoxa* Korshikov), or embryophytes (*Physcomitrella patens* (Hedw.) Bruch & Schimp., *Arabidopsis thaliana* (L.) Heynh., and *Solanum tuberosum* L.) [55]. *Chlamydomonas* mitochondrial mRNAs are also 3′ polyuridylylated and polyadenylated [54–57], but this was not observed in *P. duplex*.

Circular mRNAs are a common phenomenon across the biological spectrum, but their purpose has been difficult to determine [78]. In algal mitochondria mRNA circularization has been demonstrated in *Chlamydomonas*, where it appears to create translatable mRNAs and was hypothesized to be linked to polycytidylation [56]. In *P. duplex*, full-length coding regions were found to be circularized for seven of the twelve transcripts that were analyzed. The remaining five yielded no data so we are uncertain if they form circular transcripts. Circularization for five of the transcripts coincided with tandemly repeated template-derived motifs, AU repeats, GAACGAA, and GCGUCUU, which was not reported in *C. reinhardtii*. RNA circularization among nuclear genes occurs by way of an intron-exon back-splicing mechanism involving protein factors and conserved cis-elements [79,80], and perhaps a similar mechanism involving repeated elements occurs in mitochondria. The co-incidence of poly-cytidylation and circularization in *C. reinhardtii* led to a hypothesis that the poly(C) acted as a cis-element for circularization. In *P. duplex*, only one transcript, *atp9*, had circular mRNAs with full-length coding regions and a poly(C) addition. The other two transcripts with poly(C) additions, *atp6* and *cob*, were fragments of the coding region, suggesting that the poly(C) addition is not universally linked to the creation of translatable circular mRNAs and may not act as a signal for circularization.

These data demonstrate that mitochondrial mRNA processing is conserved in the Chlorophyceae algal clade. The polycytidylation of mitochondrial mRNAs has already been shown to be conserved across the chlorophytic algae [55]. This study extends the similarities of mitochondrial mRNA processing to include the absence of a 5′ UTR, despite *P. duplex* having larger intergenic regions than *C. reinhardtii*. Our data suggest that the cleavage of tRNAs plays a pivotal role in the maturation of two mRNAs (*cob* and *atp9*) in *P. duplex*, which is consistent with mitochondrial transcripts in all systems studied to date, from *C. reinhardtii* to humans. The difference is the size of the chondriome and the lengths of the intergenic regions. We found that the removal of tRNAs in *P. duplex* creates mature 3′ ends that, at least in two cases, require no further processing other than polycytidylation. The 5′ ends, however, are further processed until no 5′ UTR remains (Figure 4). The lack of a 5′ UTR and the accompanying translation mechanism required is not only due to the compressed chondriome of *C. reinhardtii*, but also its purposeful removal prior to translation. *Pediastrum* is in the order Sphaeropleales and is a sister clade to Volvocales, which contains *Chlamydomonas* [81], so it is possible that these processing events could be limited to those two Orders; however, we hypothesize that these processes are conserved across the Chlorophyceae.

3.2. Mitochondrial Processing in C. vulgaris Resembles That Seen in Embryophytes.

We also analyzed the mRNA termini of the mitochondrial mRNAs of *C. vulgaris* and found the 5′ and 3′ UTRs to be much longer and more variable than those observed in

P. duplex and *C. reinhardtii*. We also detected non-template polyadenylation of 3′ termini, the possibility of RNA secondary structures in processed 3′ UTRs, and the possibility of t-elements downstream of some 3′ termini. The *C. vulgaris* chondriome is one of the smaller among charophytic algae, 67,737 bp, with intergenic region sizes more similar to Chlorophyeae than higher plants, yet with gene content, gene synteny, and intron placement very similar to the much larger bryophyte chondriomes [62,63]. After post-transcriptional processing, we found that the 3′ and 5′ UTR lengths in *C. vulgaris* were more similar to those found in embryophytes than chlorophytes. In embryophytes, 5′ termini are of varying lengths, ranging from dozens to thousands of nucleotides in length, presumably due to a stepwise 5′ maturation process [29–36,82–86]. Among embryophytes, 3′ UTR lengths tend to be more consistent in length and mature mRNAs lack non-template oligonucleotide additions other than polyadenylation, which marks the mRNA for degradation [44–47]. Embryophyte mitochondria utilize tRNAs and RNA secondary structure (t-elements) to guide endonucleolytic cleavage [31–33]. The presence of relatively long 5′ and 3′ UTRs, the polyadenylation of some 3′ termini, the possibility that secondary structures may guide the formation of some 3′ termini, and the possibility that secondary structures may form in the 3′ UTRs of *C. vulgaris* mitochondrial mRNAs to presumably provide a target for 3′ specific RPFs demonstrates that their RNA processing is more similar to embryophytes than chlorophytes. One recent proteomic survey of *Arabidopsis* mitochondria estimated that 14.9% of the nuclear encoded mitochondrial-targeted proteome is devoted to mRNA processing [87]. Our results suggest that the RNA processing mechanisms and perhaps the large numbers of nuclear-encoded RPFs that locate to the mitochondria in higher plants were present in the common ancestor of all streptophytes.

In *C. vulgaris*, circularized full length coding regions were detected for five genes and only fragments were found for four other genes. Unlike *P. duplex*, there were no repeat motifs associated with the ligation sites and no polyA additions detected in the circularized transcripts. Among embryophyte mitochondria, circularized RNAs were characterized in *Hordeum vulgare* and *Arabidopsis*, but all of the examples were fragmented coding regions [88,89]. Our data suggest that the circularization of mRNAs in *C. vulgaris* may be a hybrid of chlorophytes and embryophytes.

4. Materials and Methods

4.1. Cultures

The *P. duplex* strain used in this study was morphologically identified and its chondriome sequenced and archived in GenBank (MK895949), as described in [58]. It has been maintained in Bold's basal medium [90] in an Erlenmeyer flask on a lab bench with ambient light (adjacent to a window). The *C. vulgaris* strain was collected from a pond on the University of Virginia's College at Wise campus in 2015 and cultured in an aquarium containing mud substrate collected from the same pond. This culture has been maintained in a greenhouse and supplemented with commercial plant fertilizer [59]. Its chondriome has been fully sequenced and is nearly identical to the one archived by Turmel et al. [62] (NC-005255).

4.2. RNA Extraction and CircRT-PCR

To promote RNA production, *P. duplex* cultures were incubated in a shaking incubator (50 rpm) at constant temperature (28 °C) and artificial light (500 µE m^{-2} s^{-1}) for several hours prior to extraction. Cells were pelleted using a Beckman-Coulter Avanti JXN-30 centrifuge, and pellets were resuspended in Qiagen's RNeasy extraction buffer (Germantown, MD, USA) and transferred to a bead beater tube. Cells were lysed by bead beating on a vortexer for five minutes. The lysate was then taken through the remaining Qiagen RNeasy protocol with the optional DNase step. *C. vulgaris* tissues were flash frozen in liquid nitrogen and ground with a mortar and pestle followed by RNA extraction using a Qiagen RNeasy kit, as described in Cahoon et al. [59]. All RNA samples were quantified using a NanoDrop Lite (Thermo-Fisher, Waltham, MA, USA) and stored at −80 °C.

Primers for cDNA synthesis and PCR (Supplemental Table S1) were designed using Primer3 (https://primer3.org) and synthesized by Integrated DNA Technologies (Coralville, IA, USA). RT-PCR was completed as described in Meade et al. [91]. Briefly, RNAs were artificially circularized using 2 µg of total RNA and T4 RNA ligase (New England Biolabs, Ipswich, MA, USA). cDNAs were synthesized from the circularized RNAs using the R1 primers and MMLV Reverse Transcriptase (Promega, Madison, WI, USA). These cDNAs were then used as template, along with the R1 and L1 primers to PCR amplify the 3′–5′ junctions of each transcript using Phusion DNA polymerase (ThermoFisher, Waltham, MA, USA). The products of these PCR reactions were diluted 10-fold and used as template for a second round of PCR using primers R2 and L2. This process was completed twice with RNA extracted at two different times to represent independent replicates. Naturally circularized mRNAs were detected producing cDNA directly from 2 µg of total RNA without T4 ligase treatment followed by two rounds of PCR as described above. The naturally circularized mRNA process was also completed twice.

4.3. Sequencing and Analysis

The independent replicates of the secondary PCR products for both *P. duplex and C. vulgaris* were deep sequenced, separately, using Genewiz's Amplicon EZ Illumina MiSeq service (South Plainfield, NJ, USA). Sequences were analyzed using Geneious Prime software (Biomatters, Auckland, New Zealand). Initially, sequences were matched to each gene using the Map-to-Reference function. These sequences were visually inspected, for 3′ poly-nucleotide additions, and sequence motifs. The 3′–5′ junction sites were identified using NCBI's BLAST align two or more sequences feature (https://blast.ncbi.nlm.nih.gov/Blast.cgi).

For PacBio Iso-Seq, ~1 g of total RNA from each species was polyadenylated using Lucigen's (Middleton, WI, USA) Poly(A) Polymerase Tailing kit according to the manufacturer's protocol. The tailed samples were cleaned using Qiagen's RNeasy kit. Samples were shipped to GeneWiz on dry ice for PacBio library preparation and sequencing. This process was completed once. Reads were aligned to the *P. duplex* and *C. vulgaris* chondriomes with Geneious Prime software using the Map-to-Reference function.

Logo plots were generated using the web-based service https://weblogo.berkeley.edu/logo.cgi. RNA secondary structures were predicted using RNAfold (http://rna.tbi.univie.ac.at//cgi-bin/RNAWebSuite/RNAfold.cgi).

5. Conclusions

We present evidence of mitochondrial mRNA processing from two green algae from the chlorophyte and streptophyte lineages with similarly sized circular mitochondrial genomes. The primary differences were the absence of secondary structures (t-elements) in the *P. duplex* chondriome that are important for processing in streptophytes, and the removal of the 5′ UTR in chlorophytes but not streptophytes. We hypothesize that t-elements were gained and the 5′ UTR processing lost in the common ancestor of the streptophytic algae. We also confirm the polycytidylation of the 3′ termini of *P. duplex* but not *C. vulgaris*, which is consistent with the theory that poly(C) addition is limited to chlorophytes.

Supplementary Materials: The following are available online at https://www.mdpi.com/2223-7747/10/3/576/s1, Figure S1: Logo plots, Figure S2: Hypothetical RNA secondary structures, Table S1: DNA oligo primers.

Author Contributions: Conceptualization, A.B.C., K.M.M., G.C.R.P., M.J.M.; methodology, A.B.C., K.M.M., G.C.R.P., M.J.M.; investigation, G.C.R.P., M.J.M.; resources, A.B.C., K.M.M.; writing—original draft preparation, A.B.C., G.C.R.P., M.J.M.; writing—review and editing, A.B.C., K.M.M., G.C.R.P., M.J.M.; visualization, A.B.C., G.C.R.P.; supervision, A.B.C., K.M.M.; project administration, A.B.C. All authors have read and agreed to the published version of the manuscript.

Funding: This research received no external funding.

Acknowledgments: This research was funded by the Buchanan Chair of Biology Endowment at UVA-Wise. MJM and GCRP were supported by the Fellowship in Natural Sciences program at UVA-Wise.

Conflicts of Interest: The authors declare no conflict of interest.

References

1. Rose, R.J.; Tiew, T.W.-Y.; Dashek, W. The mitochondrion. In *Plant Cells and Their Organelles*; Dashek, W.V., Miglani, G.S., Eds.; John Wiley & Sons, Ltd.: Hoboken, NJ, USA, 2017.
2. Archibald, J.M. Endosymbiosis and eukaryotic cell evolution. *Curr. Biol.* **2015**, *25*, R911–R921. [CrossRef]
3. López-García, P.; Eme, L.; Moreira, D. Symbiosis in eukaryotic evolution. *J. Theor. Biol.* **2017**, *434*, 20–33. [CrossRef] [PubMed]
4. Wang, Z.; Wu, M. An integrated phylogenomic approach toward pinpointing the origin of mitochondria. *Sci. Rep.* **2015**, *5*, 7949. [CrossRef] [PubMed]
5. Johnston, I.G.; Williams, B.P. Evolutionary inference across eukaryotes identifies specific pressures favoring mitochondrial gene retention. *Cell Syst.* **2016**, *2*, 101–111. [CrossRef] [PubMed]
6. D'Souza, A.R.; Minczuk, M. Mitochondrial transcription and translation: Overview. *Essays Biochem.* **2018**, *62*, 309–320. [CrossRef]
7. Tiranti, V.; Savoia, A.; Forti, F.; D'Apolito, M.F.; Centra, M.; Rocchi, M.; Zeviani, M. Identification of the gene encoding the human mitochondrial RNApolymerase (h-mtRPOL) by cyberscreening of the expressed sequence tags database. *Hum. Mol. Genet.* **1997**, *6*, 615–625. [CrossRef]
8. Falkenberg, M.; Gaspari, M.; Rantanen, A.; Trifunovic, A.; Larsson, N.G.; Gustafsson, C.M. Mitochondrial transcription factors B1 and B2 activate transcription of human mtDNA. *Nat. Genet.* **2002**, *31*, 289–294. [CrossRef]
9. Kanki, T.; Nakayama, H.; Sasaki, N.; Takio, K.; Alam, T.I.; Hamasaki, N.; Kang, D. Mitochondrial nucleoid and transcription factor A. *Ann. N. Y. Acad. Sci.* **2004**, *1011*, 61–68. [CrossRef]
10. Montoya, J.; Christianson, T.; Levens, D.; Rabinowitz, M.; Attardi, G. Identification of initiation sites for heavy-strand and light-strand transcription in human mitochondrial DNA. *Proc. Natl. Acad. Sci. USA* **1982**, *79*, 7195–7199. [CrossRef]
11. Montoya, J.; Gaines, G.L.; Attardi, G. The pattern of transcription of the human mitochondrial rRNA genes reveals two overlapping transcription units. *Cell* **1983**, *34*, 151–159. [CrossRef]
12. Chang, D.D.; Clayton, D.A. Precise identification of individual promoters for transcription of each strand of human mitochondrial DNA. *Cell* **1984**, *36*, 635–643. [CrossRef]
13. Gao, S.; Tian, X.; Chang, H.; Sun, Y.; Wu, Z.; Cheng, Z.; Dong, P.; Zhao, Q.; Ruan, J.; Bu, W. Two novel lncRNAs discovered in human mitochondrial DNA using PacBio full-length transcriptome data. *Mitochondria* **2018**, *38*, 41–47. [CrossRef] [PubMed]
14. Minczuk, M.; He, J.; Duch, A.M.; Ettema, T.J.; Chlebowski, A.; Dzionek, K.; Nijtmans, L.G.J.; Huynen, M.A.; Holt, I.J. TEFM (c17orf42) is necessary for transcription of human mtDNA. *Nucleic Acids Res.* **2011**, *39*, 4284–4299. [CrossRef] [PubMed]
15. Ojala, D.; Montoya, J.; Attardi, G. tRNA punctuation model of RNA processing inhuman mitochondria. *Nature* **1981**, *290*, 470–474. [CrossRef]
16. Montoya, J.; Ojala, D.; Attardi, G. Distinctive features of the 5′-terminal sequences of the human mitochondrial mRNAs. *Nature* **1981**, *290*, 465–470. [CrossRef]
17. Temperley, R.J.; Wydro, M.; Lightowlers, R.N.; Chrzanowska-Lightowlers, Z.M. Human mitochondrial mRNAs—Like members of all families, similar but different. *Biochim. Biophys. Acta* **2010**, *1797*, 1081–1085. [CrossRef]
18. Kummer, E.; Leibundgut, M.; Rackham, O.; Lee, R.G.; Boehringer, D.; Filipovska, A.; Ban, N. Unique features of mammalian mitochondrial translation initiation revealed by cryo-EM. *Nature* **2018**, *560*, 263–267. [CrossRef] [PubMed]
19. Nagaike, T.; Suzuki, T.; Katoh, T.; Ueda, T. Human mitochondrial mRNAs are stabilized with polyaenylation regulated by mitochondria-specific poly(A) polymerase and polynucleotide phosphorylase. *J. Biol. Chem.* **2005**, *280*, 19721–19727. [CrossRef]
20. Slomovic, S.; Lauer, D.; Geiger, D.; Schuster, G. Polyadenylation and degradation of human mitochondrial RNA: The prokaryotic past leaves its mark. *Mol. Cell. Biol.* **2005**, *25*, 6427–6435. [CrossRef] [PubMed]
21. Kuznetsova, I.; Siira, S.J.; Shearwood, A.-M.J.; Ermer, J.A.; Filipovska, A.; Rackham, O. Simultaneous processing and degradation of mitochondrial RNAs revealed by circularized RNA sequencing. *Nucleic Acids Res.* **2017**, *45*, 5487–5500. [CrossRef] [PubMed]
22. Mance, L.G.; Mawla, I.; Shell, S.M.; Cahoon, A.B. Mitochondrial mRNA fragments are circularized in a human HEK cell line. *Mitochondrion* **2020**, *51*, 1–6. [CrossRef]
23. Gualberto, J.M.; Newton, K.J. Plant mitochondrial genomes: Dynamics and mechanisms of mutation. *Annu. Rev. Plant Biol.* **2017**, *68*, 225–252. [CrossRef] [PubMed]
24. Hedtke, B.; Börner, T.; Weihe, A. One RNA polymerase serving two genomes. *EMBO Rep.* **2000**, *1*, 435–440. [CrossRef]
25. Kühn, K.; Richter, U.; Meyer, E.H.; Delannoy, E.; de Longevialle, A.F.; O'Toole, N.; Börner, T.; Millar, A.H.; Small, I.D.; Whelan, J. Phage-type RNA polymerase RPOTmp performs gene-specific transcription in mitochondria of *Arabidopsis thaliana*. *Plant Cell* **2009**, *21*, 2762–2779. [CrossRef]
26. Roberti, M.; Polosa, P.L.; Bruni, F.; Manzari, C.; Deceglie, S.; Gadaleta, M.N.; Cantatore, P. The MTERF family proteins: Mitochondrial transcription regulators and beyond. *Biochim. Biophys. Acta* **2009**, *1787*, 303–311. [CrossRef] [PubMed]
27. Kühn, K.; Weihe, A.; Börner, T. Multiple promoters are a common feature of mitochondrial genes in *Arabidopsis*. *Nucleic Acids Res.* **2005**, *33*, 337–346. [CrossRef] [PubMed]

28. Perrin, R.; Meyer, E.H.; Zaepfel, M.; Kim, Y.J.; Mache, R.; Grienenberger, J.M.; Gualberto, J.M.; Gagliardi, D. Two exoribonucleases act sequentially to process mature 3′-ends of atp9 mRNAs in *Arabidopsis* mitochondria. *J. Biol. Chem.* **2004**, *279*, 25440–25446. [CrossRef] [PubMed]

29. Zhang, Y.; Huang, X.; Zou, J.; Liu, Y.; Lian, T.; Nian, H. Major contribution of transcription initiation to 5′-end formation of mitochondrial steady-state transcripts in maize. *RNA Biol.* **2019**, *16*, 104–117. [CrossRef]

30. Binder, S.; Stoll, K.; Stoll, B. Maturation of 5′ ends of plant mitochondrial RNAs. *Physiol. Plant* **2016**, *157*, 280–288. [CrossRef] [PubMed]

31. Forner, J.; Weber, B.; Thuss, S.; Wildum, S.; Binder, S. Mapping of mitochondrial mRNA termini in *Arabidopsis thaliana*: T-elements contribute to 5′ and 3′ end formation. *Nucleic Acids Res.* **2007**, *35*, 3676–3692. [CrossRef] [PubMed]

32. Hanic-Joyce, P.J.; Spencer, D.F.; Gray, M.W. In vitro processing of transcripts containing novel tRNA-like sequences ('t-elements') encoded by wheat mitochondrial DNA. *Plant Mol. Biol.* **1990**, *15*, 551–559. [CrossRef] [PubMed]

33. Dombrowski, S.; Brennicke, A.; Binder, S. 3′-inverted repeats in plant mitochondrial mRNAs are processing signals rather than transcription terminators. *EMBO J.* **1997**, *16*, 5069–5076. [CrossRef]

34. Gobert, A.; Gutmann, B.; Taschner, A.; GöBringer, M.; Holzmann, J.; Hartmann, R.K.; Rossmanith, W.; Giegé, P. A single *Arabidopsis* organellar protein has RNase P activity. *Nat. Struct. Mol. Biol.* **2010**, *17*, 740–744. [CrossRef] [PubMed]

35. Gutmann, B.; Gobert, A.; Giegé, P. PRORP proteins support RNase P activity in both organelles and the nucleus in *Arabidopsis*. *Genes Dev.* **2012**, *26*, 1022–1027. [CrossRef] [PubMed]

36. Stoll, B.; Binder, S. Two NYN domain containing putative nucleases are involved in transcript maturation in *Arabidopsis* mitochondria. *Plant J.* **2016**, *85*, 278–288. [CrossRef] [PubMed]

37. Jonietz, C.; Forner, J.; Hölzle, A.; Thuss, S.; Binder, S. RNA PROCESSING FACTOR2 is required for 5′ end processing of nad9 and cox3 mRNAs in mitochondria of *Arabidopsis thaliana*. *Plant Cell* **2010**, *22*, 443–453. [CrossRef]

38. Hölzle, A.; Jonietz, C.; Törjek, O.; Altmann, T.; Binder, S.; Forner, J. A RESTORER OF FERTILITY-like PPR gene is required for 5′-end processing of the nad4 mRNA in mitochondria of *Arabidopsis thaliana*. *Plant J.* **2011**, *65*, 65,737–744. [CrossRef]

39. Hauler, A.; Jonietz, C.; Stoll, B.; Stoll, K.; Braun, H.-P.; Binder, S. RNA PROCESSING FACTOR 5 is required for efficient 5′ cleavage at a processing site conserved in RNAs of three different mitochondrial genes in *Arabidopsis thaliana*. *Plant J.* **2013**, *74*, 593–604. [CrossRef]

40. Haili, N.; Arnal, N.; Quadrado, M.; Amiar, S.; Tcherkez, G.; Dahan, J.; Briozzo, P.; Colas des Francs-Small, C.; Vrielynck, N.; Mireau, H. The pentatricopeptide repeat MTSF1 protein stabilizes the nad4 mRNA in *Arabidopsis* mitochondria. *Nucleic Acids Res* **2013**, *41*, 6650–6663. [CrossRef]

41. Hammani, K.; Giegé, P. RNA metabolism in plant mitochondria. *Trends Plant Sci.* **2014**, *19*, 380–389. [CrossRef]

42. Brown, G.G.; des Francs-Small, C.C.; Ostersetzer-Biran, O. Group II intron splicing factors in plant mitochondria. *Front. Plant Sci.* **2014**, *5*, 35. [CrossRef] [PubMed]

43. Small, I.D.; Schallenberg-Rüdinger, M.; Takenaka, M.; Mireau, H.; Ostersetzer-Biran, O. Plant organellar RNA editing: What 30 years of research has revealed. *Plant J.* **2020**, *101*, 1040–1056. [CrossRef] [PubMed]

44. Gagliardi, D.; Leaver, C.J. Polyadenylation accelerates the degradation of the mitochondrial mRNA associated with cytoplasmic male sterility in sunflower. *EMBO J.* **1999**, *18*, 3757–3766. [CrossRef] [PubMed]

45. Lupold, D.S.; Caoile, A.G.; Stern, D.B. Polyadenylation occurs at multiple sites in maize mitochondrial cox2 mRNA and is independent of editing status. *Plant Cell* **1999**, *11*, 1565–1578. [CrossRef] [PubMed]

46. Kühn, J.; Tengler, U.; Binder, S. Transcript lifetime is balanced between stabilizing stem-loop structures and degradation-promoting polyadenylation in plant mitochondria. *Mol. Cell. Biol.* **2001**, *21*, 731–742. [CrossRef] [PubMed]

47. Schuster, G.; Stern, D. RNA polyadenylation and decay in mitochondria and chloroplasts. *Prog. Mol. Biol. Transl. Sci.* **2009**, *85*, 393–422. [CrossRef]

48. Harris, E.H. *The Chlamydomonas Sourcebook*, 2nd ed.; Harris, E.H., Stern, D.B., Witman, G.B., Eds.; Academic Press: Oxford, UK, 2009.

49. Gray, M.W.; Boer, P.H. Organization and expression of algal (*Chlamydomonas reinhardtii*) mitochondrial DNA. *Phil. Trans. R. Soc. Lon. B Biol. Sci.* **1988**, *319*, 135–147. [CrossRef]

50. Laflamme, M.; Lee, R.W. Mitochondrial genome conformation among CW-group chlorophycean algae. *J. Phycol.* **2003**, *39*, 213–220. [CrossRef]

51. Mallet, M.A.; Lee, R.W. Identification of three distinct *Polytomella* lineages based on mitochondrial DNA features. *J. Eukaryot. Microbiol.* **2006**, *53*, 79–84. [CrossRef]

52. Tracy, R.L.; Stern, D.B. Mitochondrial transcription initiation: Promoter structures and RNA poymerases. *Curr. Genet.* **1995**, *28*, 205–216. [CrossRef]

53. Duby, F.; Cardol, P.; Matgne, R.F.; Remacle, C. Structure of the telomeric ends of mt DNA, transcriptional analysis and complex I assembly in the dum24 mitochondrial mutant of *Chlamydomonas reinhardtii*. *Mol. Genet. Genom.* **2001**, *266*, 109–114. [CrossRef]

54. Zimmer, S.L.; Schein, A.; Zipor, G.; Stern, D.B.; Schuster, G. Polyadenylation in *Arabidopsis* and *Chlamydomonas* organelles: The input of nucleotidyltransferases, poly(A) polymerases and polynucleotide phosphorylase. *Plant J.* **2009**, *59*, 88–99. [CrossRef] [PubMed]

55. Salinas-Giegé, T.; Cavaiuolol, M.; Cognat, V.; Ubrig, E.; Remacle, C.; Duchêne, A.-M.; Vallon, O.; Maréchal-Drouard, L. Poly-cytidylation of mitochondrial mRNAs in *Chlamydomonas reinhardtii*. *Nucleic Acids Res.* **2017**, *45*, 12963–12973. [CrossRef]

56. Cahoon, A.B.; Qureshi, A.A. Leaderless mRNAs are circularized in *Chlamydomonas reinhardtii* mitochondria. *Curr. Genet.* **2018**, *64*, 1321–1333. [CrossRef]

57. Gallaher, S.D.; Fitz-Gibbon, S.T.; Strenkert, D.; Purvine, S.O.; Pellegrini, M.; Merchant, S.S. High-throughput sequencing of the chloroplast and mitochondrion of *Chlamydomonas reinhardtii* to generate improved de novo assemblies, analyze expression patterns and transcript speciation, and evaluate diversity among laboratory strains and wild isolates. *Plant J.* **2017**, *93*, 545–565. [CrossRef]

58. Levy, S.; Schuster, G. Polyadenylation and degradation of RNA in the mitochondria. *Biochem. Soc. Trans.* **2016**, *44*, 1475–1482. [CrossRef]

59. Cahoon, A.B.; Nauss, J.A.; Stanley, C.D.; Qureshi, A. Deep Transcriptome Sequencing of Two Green Algae, *Chara vulgaris* and *Chlamydomonas reinhardtii*, Provides no Evidence of Organellar RNA Editing. *Genes* **2017**, *8*, 80. [CrossRef] [PubMed]

60. Farwagi, A.; Fučíková, K.; McManus, H.A. Phylogenetic patterns of gene rearrangements in four mitochondrial genomes from the green algal family Hydrodictyaceae (Sphaeropleales, Chlorophyceae). *BMC Genom.* **2015**, *16*, 826. [CrossRef] [PubMed]

61. Proulex, G.C.R.; Lor, B.; Manoylov, K.M.; Cahoon, A.B. The chloroplast and mitochondrial genomes of the green alga *Pediastrum duplex* isolated from central Georgia (USA). *Mitochondrial DNA Part B Resour.* **2019**, *4*, 3070–3071. [CrossRef]

62. Turmel, M.; Otis, C.; Lemieux, C. The mitochondrial genome of *Chara vulgaris*: Insights into the mitochondrial DNA architecture or the last common ancestor of green algae and land plants. *Plant Cell* **2003**, *15*, 1888–1903. [CrossRef] [PubMed]

63. Turmel, M.; Otis, C.; Lemieux, C. Tracing the evolution of streptophyte algae and their mitochondrial genome. *Genome Biol. Evol.* **2013**, *5*, 1817–1835. [CrossRef] [PubMed]

64. Greiner, S.; Lehwark, P.; Bock, R. OrganellarGenomeDraw (OGDRAW) version 1.3.1: Expanded toolkit for the graphical visualization of organellar genomes. *Nucleic Acids Res.* **2019**, *47*, W59–W64. [CrossRef]

65. Leliaert, F.; Smith, D.R.; Moreau, H.; Herron, M.D.; Verbruggen, H.; Delwiche, C.F.; De Clerck, O. Phylogeny and molecular evolution of the green algae. *Crit. Rev. Plant Sci.* **2012**, *31*, 1–46. [CrossRef]

66. Boer, P.F.; Bonen, L.; Lee, R.W.; Gray, M.W. Genes for respiratory chain proteins and ribosomal RNAs are present on a 16-kilobase-pair DNA species from *Chlamydomonas reinhardtii* mitochondria. *Proc. Natl. Acad. Sci. USA* **1985**, 3340–3344. [CrossRef] [PubMed]

67. Turmel, M.; Lemieux, C.; Burger, G.; Lang, B.F.; Otis, C.; Plante, I.; Gray, M.W. The complete mitochondrial DNA sequences of *Nephroselmis olivacea* and *Pedinomonas minor*: Two radically different evolutoionary patterns within green algae. *Plant Cell* **1999**, *11*, 1717–1729. [CrossRef] [PubMed]

68. Nedelcu, A.M.; Lee, R.W.; Lemieux, C.; Gray, M.W.; Burger, G. The complete mitochondrial DNA sequence of *Scenedesmus obliquus* reflects an intermediate stage in the evolution of the green algal mitochondrial genome. *Genome Res.* **2000**, *10*, 819–831. [CrossRef] [PubMed]

69. Robbens, S.; Derelle, E.; Ferraz, C.; Wuyts, J.; Moreau, H.; Van de Peer, Y. The complete chloroplast and mitochondrial DNA sequences of *Ostreococcus tauri*: Organelle genomes of the smallest eukaryote are examples of compaction. *Mol. Biol. Evol.* **2007**, *24*, 956–968. [CrossRef] [PubMed]

70. Brouard, J.-S.; Otis, C.; Lemieux, C.; Turmel, M. Chloroplast DNA sequence of the green alga *Oedogonium cardiacum* (Chlorophyceae): Unique genome architecture, derived characters shared with the Chaetophorales and novel genes acquired through horizontal transfer. *BMC Genom.* **2008**, *9*, 290. [CrossRef]

71. Pombert, J.-F.; Otis, C.; Lemieux, C.; Turmel, M. The complete mitochondrial DNA sequence of the green alga *Pseudenclonium akinetum* (Ulvophyceae) highlights distinctive evolutionary trends in the Chlorophyta and suggests a sister-group relationship between the Ulvophyceae and Chlorophyceae. *Mol. Biol. Evol.* **2004**, *21*, 922–935. [CrossRef]

72. Pombert, J.-F.; Otis, C.; Turmel, M.; Lemieux, C. The mitochondrial genome of the prasinophyte *Prasinoderma colonial* reveals two *trans*-spliced group I introns in the large subunit rRNA gene. *PLoS ONE* **2013**, *8*, e84325. [CrossRef]

73. Smith, D.R.; Hamaji, T.; Olson, B.J.S.C.; Durand, P.M.; Ferris, P.; Michod, R.E.; Featherston, J.; Nozaki, H.; Keeling, P.J. Organelle genome complexity scales positively with organism size in volvocine green algae. *Mol. Biol. Evol.* **2013**, *30*, 793–797. [CrossRef] [PubMed]

74. Jeong, H.; Lim, J.-M.; Park, J.; Sim, Y.M.; Choi, H.-G.; Lee, J.; Jeong, W.-J. Plastid and mitochondrion genomic sequences from arctic *Chlorella* sp. ArM0029B. *BMC Genom.* **2014**, *15*, 286. [CrossRef] [PubMed]

75. Fučíková, K.; Lewis, P.O.; González-Halphen, D.; Lewis, L.A. Gene arrangement convergence, diverse intron content, and genetic code modifications in mitochondrial genomes of Sphaeropleales (Chlorophyta). *Genome Biol. Evol.* **2014**, *6*, 2170–2180. [CrossRef]

76. Nishiyama, T.; Sakayama, H.; de Vries, J.; Buschmann, H.; Saint-Marcoux, D.; Ullrich, K.K.; Haas, F.B.; Vanderstraeten, L.; Becker, D.; Lang, D.; et al. The *Chara* genome: Secondary complexity and implications for plant terrestrialization. *Cell* **2018**, *174*, 448–464. [CrossRef]

77. Karol, K.G.; McCourt, R.M.; Cimino, M.T.; Delwiche, C.F. The closest living relatives of land plants. *Science* **2001**, *294*, 2351–2353. [CrossRef] [PubMed]

78. Guria, A.; Sharma, P.; Natesan, S.; Pandi, G. Circular RNAs—The road less traveled. *Front. Mol. Biosci.* **2020**, *6*, 146. [CrossRef]

79. Starke, S.; Jost, I.; Rossbach, O.; Schneider, T.; Schriner, S.; Hung, L.-H.; Bindereif, A. Exon circularization requires canonical splice signals. *Cell Rep.* **2015**, *10*, 103–111. [CrossRef] [PubMed]

80. Bolha, L.; Ravnik-Glavač, M.; Glavač, D. Circular RNAs: Biogenesis, Function and a Role as Possible Cancer Biomarkers. *Int. J. Genom.* **2017**, *2017*, 6218353. [CrossRef]

81. Tippery, N.P.; Fučiková, K.; Lewis, P.O.; Lewis, L.A. Probing the monophyly of the Sphaeropleales (Chlorophyceae) using data from five genes. *J. Phycol.* **2012**, *48*, 1482–1493. [CrossRef]
82. Arnal, N.; Quadrado, M.; Simon, M.; Mireau, H. A restorer-of-fertility like pentatricopeptide repeat gene directs ribonucleolytic processing within the coding sequence of rps3-rpl16 and orf240a mitochondrial transcripts in *Arabidopsis thaliana*. *Plant J.* **2014**, *78*, 134–145. [CrossRef]
83. Jonietz, C.; Forner, J.; Hildebrandt, T.; Binder, S. RNA PROCESSING FACTOR 3 is crucial for the accumulation of mature ccmC transcripts in mitochondria of *Arabidopsis thaliana* accession Columbia. *Plant Physiol.* **2011**, *157*, 1430–1439. [CrossRef] [PubMed]
84. Stoll, B.; Stoll, K.; Steinhilber, J.; Jonietz, C.; Binder, S. Mitochondrial transcript length polymorphisms are a widespread phenomenon in *Arabidopsis thaliana*. *Plant Mol. Biol.* **2012**, *1*, 221–233. [CrossRef] [PubMed]
85. Stoll, B.; Zendler, D.; Binder, S. RNA processing factor 7 and polynucleotide phosphorylase are necessary for processing and stability of nad2 mRNA in *Arabidopsis* mitochondria. *RNA Biol.* **2014**, *11*, 968–976. [CrossRef] [PubMed]
86. Stoll, K.; Jonietz, C.; Binder, S. In *Arabidopsis thaliana* two co-adapted cyto-nuclear systems correlate with distinct ccmC transcript sizes. *Plant J.* **2015**, *81*, 247–257. [CrossRef]
87. Fuchs, P.; Rugen, N.; Carrie, C.; Elsässer, M.; Finkemeier, I.; Giese, J.; Hildebrandt, T.M.; Kühn, K.; Maurino, V.G.; Ruberti, C.; et al. Single organelle function and organization as estimated from *Arabidopsis* mitochondrial proteomics. *Plant J.* **2019**, *101*, 420–441. [CrossRef]
88. Darbani, B.; Noeparvar, S.; Borg, S. Identification of circular RNAs from parental genes involved in multiple aspects of cellular metabolism in barley. *Front. Plant Sci.* **2016**, *7*, 776. [CrossRef]
89. Sun, X.; Wang, X.; Ding, J.; Wang, Y.; Wang, J.; Zhang, X.; Che, Y.; Liu, Z.; Zhang, X.; Ye, J.; et al. Integrative analysis of *Arabidopsis thaliana* transcriptomics reveals intuitive splicing mechanism for circular RNA. *FEBS Lett.* **2016**, *590*, 3510–3516. [CrossRef]
90. Bischoff, H.W.; Bold, H.C. Phycological studies. IV. Some soil algae from Enchanted Rock and related algal species. *Univ. Tex. Publ.* **1963**, *6318*, 1–95.
91. Meade, M.J.; Proulex, G.C.R.; Manoylov, K.M.; Cahoon, A.B. Chloroplast MRNAs are 3′ Polyuridylylated in The Green Alga *Pithophora roettleri* (Cladophorales). *J. Phycol.* **2020**, *56*, 1124–1134. [CrossRef]

Article

Comparative Analysis of Chloroplast Genomes of Four Medicinal Capparaceae Species: Genome Structures, Phylogenetic Relationships and Adaptive Evolution

Dhafer A. Alzahrani [1,*], Enas J. Albokhari [1,2,*], Samaila S. Yaradua [1] and Abidina Abba [1]

[1] Department of Biological Sciences, Faculty of Sciences, King Abdulaziz University, P.O. Box 80203, Jeddah 21589, Saudi Arabia; dryaradua@gmail.com (S.S.Y.); abidin2007@gmail.com (A.A.)

[2] Department of Biological Sciences, Faculty of Applied Sciences, Umm Al-Qura University, Makkah 24351, Saudi Arabia

* Correspondence: dalzahrani@kau.edu.sa (D.A.A.); enasbokhary@hotmail.com (E.J.A.); Tel.: +966-555546409 (D.A.A.)

Citation: Alzahrani, D.A.; Albokhari, E.J.; Yaradua, S.S.; Abba, A. Comparative Analysis of Chloroplast Genomes of Four Medicinal Capparaceae Species: Genome Structures, Phylogenetic Relationships and Adaptive Evolution. *Plants* **2021**, *10*, 1229. https://doi.org/10.3390/plants10061229

Academic Editors: Nunzia Scotti and Rachele Tamburino

Received: 4 March 2021
Accepted: 31 May 2021
Published: 17 June 2021

Abstract: This study presents for the first time the complete chloroplast genomes of four medicinal species in the Capparaceae family belonging to two different genera, *Cadaba* and *Maerua* (i.e., *C. farinosa*, *C. glandulosa*, *M. crassifolia* and *M. oblongifolia*), to investigate their evolutionary process and to infer their phylogenetic positions. The four species are considered important medicinal plants, and are used in the treatment of many diseases. In the genus *Cadaba*, the chloroplast genome ranges from 156,481 bp to 156,560 bp, while that of *Maerua* ranges from 155,685 bp to 155,436 bp. The chloroplast genome of *C. farinosa*, *M. crassifolia* and *M. oblongifolia* contains 138 genes, while that of *C. glandulosa* contains 137 genes, comprising 81 protein-coding genes, 31 tRNA genes and 4 rRNA genes. Out of the total genes, 116–117 are unique, while the remaining 19 are replicated in inverted repeat regions. The *psbG* gene, which encodes for subunit K of NADH dehydrogenase, is absent in *C. glandulosa*. A total of 249 microsatellites were found in the chloroplast genome of *C. farinosa*, 251 in *C. glandulosa*, 227 in *M. crassifolia* and 233 in *M. oblongifolia*, the majority of which are mononucleotides A/T found in the intergenic spacer. Comparative analysis revealed variable hotspot regions (*atpF*, *rpoC2*, *rps19* and *ycf1*), which can be used as molecular markers for species authentication and as regions for inferring phylogenetic relationships among them, as well as for evolutionary studies. The monophyly of Capparaceae and other families under Brassicales, as well as the phylogenetic positions of the studied species, are highly supported by all the relationships in the phylogenetic tree. The cp genomes reported in this study will provide resources for studying the genetic diversity of Capparaceae, as well as resolving phylogenetic relationships within the family.

Keywords: Capparaceae; chloroplast genome; *Cadaba*; *Maerua*; phylogenetic relationships

1. Introduction

The family Capparaceae, whose members are distributed in both arid and semi-arid areas, has about 470 morphologically diverse species in ca. 17 genera, which include *Cadaba* and *Maerua* [1–3]. Members of the family possess highly essential compounds used in folk medicine that are extracted from them [4]. The four species in question are considered important medicinal plants and are used in the treatment of many diseases. Most *Cadaba* species contain important compounds, such as alkaloids, sesquiterpene lactones and cadabicine. *Cadaba farinosa* and *Cadaba glandulosa* are used as purgative, anthelmintic, antisyphilitic, emmenagogue, aperient, antiscorbutic, and antiphlogistic substances; for liver damage and cancer, dysentery, fever and body pain; in therapy as a hepatoprotective and hypoglycemic [5,6]. *Maerua crassifolia* and *Maerua oblongifolia* species are used in the treatment of fever, stomach troubles, skin infections, diabetes mellitus, epilepsy and abdominal colic; they demonstrate antimicrobial and antioxidant properties and are used

105

for hypocholesterolemia, wound-healing, intestinal disorders like abdominal cramps and hookworms, anthrax, severe mumps, tetanus and eye disease [5,7–11].

According to the taxonomic status of Capparaceae, Capparideae was placed under the cohort Parietales [12]. Later, Capparidaceae and Cruciferae were placed under suborder Capparidineae, order Rhoedales [13], and Capparidaceae was classified under Capparidales [14,15]. After some decades, Capparaceae was placed under Capparales [1,16], and finally under order Brassicales [17–20]. Previous studies, with the exception of Hutchinson [14], reported that Brassicaceae and Capparaceae are sister taxa [1,21–30]. The two families (Brassicaceae and Capparaceae) are considered as one family—Brassicaceae sensu lato (s.l.)—by some authors [17,18,31,32]. Phylogenetic relationship studies using genes from chloroplast and nuclear genomes [29,33] confirmed the monophyly of Brassicaceae and Capparaceae. Within Capparaceae, there are two subfamilies, Cleomoideae and Capparoideae; these subfamilies are elevated to family by some studies of Brassicales [14,34]. Currently, as adopted by the Angiosperm Phylogeny Group [19,20], Cleomaceae, Capparaceae and Brassicaceae are considered as a single family.

There has been some shifting of a few genera between the families Brassicaceae and Capparaceae, such as two genera, *Dipterygium* and *Puccionia*, previously belonging to Brassicaceae [14] being moved to Capparaceae under the subfamily Dipterigioideae, based on the presence of methyl-glucosinolate [35,36]. The genus *Stixis* L. was removed from the Capparaceae family and represents as a new family, which is called Stixaceae Doweld (including the genus *Forchhammeria* Lieb.), yet, it is still considered under Brassicaceae sensu lato, excluding *Forchhammeria*, as it is more closely related to Resedaceae than Brassicaceae [37].

Genetic information is a reliable means of understanding evolutionary relationships among species in various taxonomic hierarchies. The genetic information in the chloroplast genome contains sufficient information for comparison analysis and studies of species diversification, due to the presence of functional genes that have a vital role in plant cells [38]. The chloroplast organelle functions in carbon fixation and photosynthesis in plants [39]. The chloroplast genome is more conserved than other genomes found in plants. Generally, the chloroplast genome is circular, double-stranded and has a quadripartite structure, including a large single copy (LSC), as well as a small single copy (SSC) and a pair of repeats (IRa and IRb) [40]. The chloroplast genome is uniparentally inherited, and this characteristic makes it highly conserved in structure and gene content [41,42]. However, different kinds of mutations do occur [43], which consequently lead to sequence divergence among species and could be used to study evolutionary relationships in plants [44]. Despite the importance of the plastome in modern taxonomy, chloroplast genomes of only three species in the whole Capparaceae family, including three varieties, have been reported: *Capparis spinosa* [45], *Capparis spinosa* var. *spinosa*, *Capparis spinosa* var. *herbacea*, *Capparis spinosa* var. *ovata* [46] and *Capparis decidua*.

This study obtained the first complete chloroplast genome of the genus *Cadaba* (*Cadaba farinosa* and *Cadaba glandulosa*) and genus *Maerua* (*Maerua crassifolia* and *Maerua oblongifolia*) using Illumina sequencing technology. This study also analyzed and compared the features of the cp genomes to provide resources of genetic data for the four species. We reconstructed the phylogenetic relationship between Capparaceae, Cleomaceae and Brassicaceae to infer the phylogenetic positions of the species within the families.

2. Results

2.1. Characteristics of Four Chloroplast Genomes

Previous studies have shown that the plastomes of flowering plants are greatly conserved in structural organization and gene content, but contraction and expansion do occur [47,48]. Each complete chloroplast genome of *C. farinosa*, *C. glandulosa*, *M. crassifolia* and *M. oblongifolia* has a circular and quadripartite structure. The genome of *C. farinosa*, *C. glandulosa*, *M. crassifolia* and *M. oblongifolia* ranged from 156,560 bp (*C. glandulosa*) to 155,436 bp (*M. oblongifolia*); the coding region ranged from 78,080 bp (*C. farinosa*) to 76,614 bp

(*C. glandulosa*), corresponding to 49.89% and 48.93% of the total genome length. The LSC regions ranged from 85,681 bp (*C. glandulosa*) to 84,153 bp (*M. oblongifolia*) in size, whereas the SSC ranged from 18,481 bp (*M. oblongifolia*) to 18,031 bp (*C. glandulosa*); the pair of inverted repeats are separated by the small single copy region and ranged from 26,430 bp (*C. farinosa*) to 26,294 bp (*M. crassifolia*) (Table 1 and Figure 1). These four Capparaceae chloroplast genome sequences were deposited in GenBank (accession numbers: *C. farinosa*, MN603027; *C. glandulosa*, MN603028; *M. crassifolia*, MN603029 and *M. oblongifolia*, MN603030).

Table 1. Base content in the *C. farinosa*, *C. glandulosa*, *M. crassifolia* and *M. oblongifolia* chloroplast genomes.

Species	*C. farinosa*	*C. glandulosa*	*M. crassifolia*	*M. oblongifolia*
Genome size (bp)	156,481	156,560	155,685	155,436
IR (bp)	26,430	26,424	26,294	26,401
LSC (bp)	85,565	85,681	84,624	84,153
SSC (bp)	18,056	18,031	18,473	18,481
Total number of genes	138	137	138	138
rRNA	4	4	4	4
tRNA	31	31	31	31
Protein-coding genes	81	80	81	81
A%	31	31	31	31
C%	18	18	18	18
T%	32	32	32	32
G%	17	17	17	17

In the four species, the plastomes are well conserved in gene order and number of genes, with slight variation in the presence of the *psbG* gene, which is absent in *C. glandulosa*. The result of the gene annotation revealed a total of 137 in *C. glandulosa* and 138 genes in *C. farinosa*, *M. crassifolia* and *M. oblongifolia*, among which 116–117 are situated in the SSC and the LSC copy regions, and 19 genes are located in the IRa and IRb. The plastome contained 80 protein-coding genes in *C. glandulosa* and 81 in other species, four rRNA genes and 31 tRNA genes (Figure 1 and Table 2). Eight protein-coding genes, four rRNA and seven tRNA were found in the IR regions. In the *C. glandulosa* species, the LSC region contained 61 protein-coding genes, whereas it included 62 in other species and 23 tRNA genes; the remaining 12 protein-coding genes and 1 tRNA are situated in the SSC region.

Some protein-coding genes and tRNA genes in the chloroplast genome of angiosperms contain an intron [49,50], as is found in the plastomes of the four species (*C. farinosa*, *C. glandulosa*, *M. crassifolia* and *M. oblongifolia*). In the total genes of the cp genomes (out of the total genes in all chloroplast genomes), 13 genes in *C. glandulosa* and *M. crassifolia* and 14 genes in *C. farinosa* and *M. oblongifolia* include an intron; some genes are protein-coding genes (nine in *C. farinosa* and *M. oblongifolia* and eight in *C. glandulosa* and *M. crassifolia*) while the remaining five are tRNA genes. Four genes (*rpl2*, *ndhB*, *trnI-GAU* and *trnA-UGC*) that have introns are situated in the inverted repeat region, *ndhA* is located in the SSC region and the remainder is found in the LSC region. All genes have only one intron and only two genes, namely *ycf3* and *clpP*, have two introns. The gene *trnK-UUU* has the longest intron of 2542–2571 bp; this is a result of the *matK* gene being located within the intron of the gene.

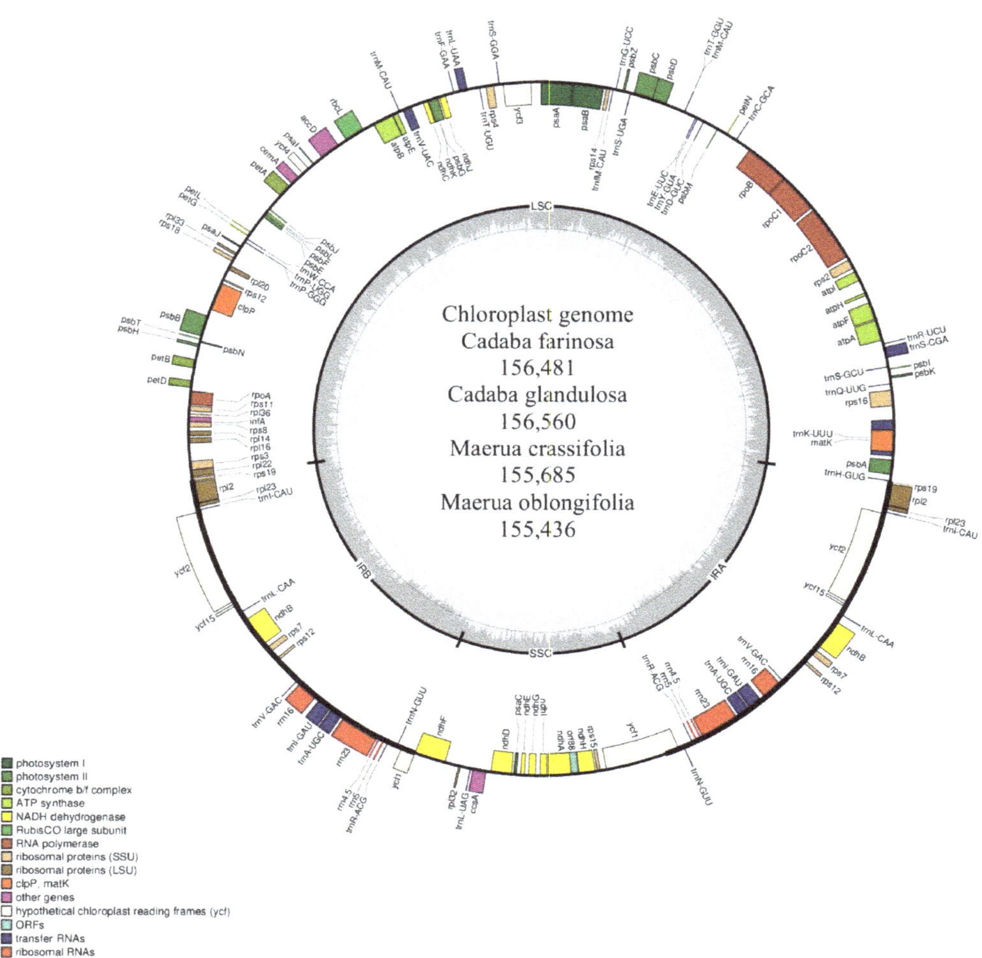

Figure 1. Chloroplast genome maps of the four Capparaceae species. Genes drawn inside the circle are transcribed clockwise, while those outside the circle are transcribed counter-clockwise. The inner dark gray circle corresponds to GC content and the inner light gray circle corresponds to the AT content. Different colors are used as a representation of distinctive genes within separate functional groups.

Table 2. Gene contents in the chloroplast genomes of *Cadaba* and *Maerua* species.

Category	Gene Groups	Gene Names
	Ribosomal RNA genes (rRNA)	*rrn5, rrn4.5, rrn16, rrn23*
RNA genes	Transfer RNA genes (tRNA)	*trnH-GUG, trnK-UUU [+], trnQ-UUG, trnS-GCU, trnS-CGA [+], trnR-UCU, trnC-GCA, trnD-GUC, trnY-GUA, trnE-UUC, trnT-GGU, trnS-UGA, trnG-GCC, trnfM-CAU, trnS-GGA, trnT-UGU, trnL-UAA [+], trnF-GAA, trnV-UAC [+], trnM-CAU, trnW-CCA, trnP-UGG, trnP-GGG, trnI-CAU [a], trnL-CAA [a], trnV-GAC [a], trnI-GAU [+,a], trnA-UGC [+,a], trnR-ACG [a], trnN-GUU [a], trnL-UAG*

Table 2. *Cont.*

Category	Gene Groups	Gene Names
Ribosomal proteins	Small ribosomal subunit	*rps2, rps3, rps4, rps7* [a]*, rps8, rps11, rps12* [a]*, rps14, rps15, rps16* [+]*, rps18, rps19*
Transcription	Large ribosomal subunit	*rpl2* [+,a]*, rpl14, rpl16, rpl20, rpl22, rpl23* [a]*, rpl32, rpl33, rpl36*
	DNA dependent RNA polymerase	*rpoA, rpoB, rpoC1* [+]*, rpoC2*
Protein-coding genes	Photosystem I	*psaA, psaB, psaC, psaI, psaJ, ycf3* [++]
	Photosystem II	*psbA, psbB, psbC, psbD, psbE, psbF, psbG, psbH, psbI, psbJ, psbK, psbL, psbM, psbN, psbT, psbZ, ndhK* [*]
	Subunit of cytochrome	*petA, petB, petD, petG, petL, petN*
	Subunit of synthase	*atpA, atpB, atpE, atpF* [+]*, atpH, atpI*
	Large subunit of Rubisco	*rbcL*
	NADH dehydrogenase	*ndhA* [+]*, ndhB* [+,a]*, ndhC, ndhD, ndhE, ndhF, ndhG, ndhH, ndhI, ndhJ*
Other genes	ATP dependent protease subunit P	*clpP* [++]
	Chloroplast envelope membrane protein	*cemA*
	Maturase	*matK*
	Subunit of acetyl-CoA carboxylase	*accD*
	C-type cytochrome synthesis	*ccsA*
	Translation initiation factor	*infA*
	Hypothetical proteins	*ycf2* [a]*, ycf4, ycf15* [a]
	Component of the TIC complex	*ycf1* [a]

+ Gene with one intron, ++ gene with two introns and a gene with multiple copies. [a] gene with multiple copies. [*] *ndhK* in group photosystem II in *C. farinosa* and group NADH dehydrogenase in *C. glandulosa, M. crassifolia* and *M. oblongifolia*.

2.2. Codon Usage

One of the factors shaping the evolution of the chloroplast genome is codon usage, which occurs as a result of bias in mutation [51]. The codon usage bias in the plastome was computed using the protein-coding genes and tRNA gene nucleotide sequences— 104,575 bp, 106,488 bp, 105,750 bp and 99,100 bp for *C. farinosa, C. glandulosa, M. crassifolia* and *M. oblongifolia*, respectively. Supplementary Tables S1–S4 present the relative synonymous codon usage (RSCU) of each codon in the genome; these results suggested that all the genes are encoded by 33,686 codons in *C. farinosa*, 34,303 codons in *C. glandulosa*, 34,064 codons in *M. crassifolia* and 31,920 codons in *M. oblongifolia*. The most frequently occurring codons are for the amino acids leucine 3290 (9.76%), 3599 (10.49%), 3452 (10.13%) and 2951 (9.24%), respectively (Figure 2), which have been found in other flowering plant genomes [52], whereas methionine has the fewest codons in the genome at 620 (1.84%) in *C. farinosa* and 606 (1.89%) in *M. oblongifolia*, and for tryptophan it is 674 (1.96%) in *C. glandulosa* and 695 (2.04%) in *M. crassifolia*. A- and T- endings were discovered to be less frequent than their counterparts G- and C-. The codon usage bias is low in the cp genomes of Capparaceae species (Tables S1–S4). Additionally, the results show that the RSCU values of 27 codons were >1, all with A/T- endings, whereas the other 28 codons were <1, and all ended with G/C. The RSCU values of tryptophan and methionine amino acids are 1, so they are the only amino acids with no codon bias.

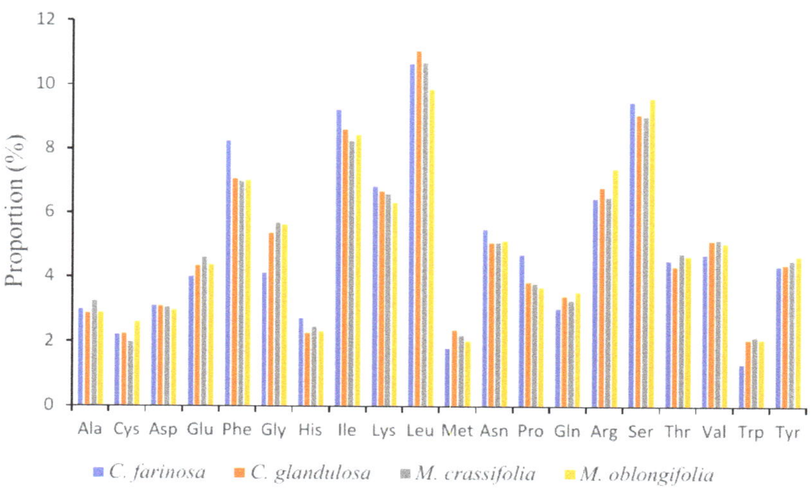

Figure 2. Amino acid frequencies in the four Capparaceae chloroplast genomes' protein-coding sequences.

2.3. RNA Editing Sites

RNA editing features a set of processes that comprise of insertion, deletion or modifications of nucleotides that alter the DNA-encoded sequence [53], which represents a way to create transcript and protein diversity [54]. Some chloroplast RNA editing sites are preserved in plants [55]. The RNA editing sites in the *C. farinosa*, *C. glandulosa*, *M. crassifolia* and *M. oblongifolia* chloroplast genomes were predicted using the PREP suite program; the first codon position of the first nucleotide was used in all of the analyses. The results show that conversion of the amino acid serine into leucine was the majority of the conversions in the codon positions (Tables S5–S8). This conversion is found to occur more frequently [56]. In total, 48 editing sites in the genus *Maerua* and 50 in the genus *Cadaba* were revealed by the program. Twenty protein-coding genes in *C. farinosa* and 19 protein-coding genes in *C. glandulosa*, *M. crassifolia* and *M. oblongifolia* were distributed across the editing sites. As stated in previous studies [57–59], the *ndhB* genes have the largest number of editing sites (nine sites), followed by *ndhD* (nine sites in *C. farinosa* and *M. oblongifolia* and eight sites in *C. glandulosa* and *M. crassifolia*), while *accD*, *atpF*, *ccsA*, *clpP*, *PsaI*, *psbG*, *psbF*, *rpoA*, *rpl20*, *rps2 and rps16* have at least one site each. Certain RNA sites, amidst all the conversions in the RNA editing (modification) sites, changed the amino acid from proline to serine. RNA-predicting sites in the first codon of the first nucleotides are not present in the following genes: *atpA, atpB, atpI, ccsA* (only in *C. glandulosa*), *petB, petD, petG, petL, psaB, psbB, psbL, rpl2, rpl20* (except in *M. oblongifolia*), *rpl23, rps8 and ycf3*, among others. This result indicated that the preservation of RNA editing is fundamental [60,61].

2.4. Repeat Analysis

2.4.1. Long Repeats

Repeat sequences in the chloroplast genomes of the four Capparaceae species were determined by the REPuter program with default settings; the obtained results clearly show that forward, reverse, palindrome and complemented repeats were detected in the cp genomes (Figure 3). The long repeat analysis in *C. farinosa*, *C. glandulosa*, *M. crassifolia* and *M. oblongifolia* showed 25–26–18–24 palindromic repeats, 12–12–14–13 forward repeats, 9–8–16–11 reverse repeats and 3–3–1–1 complement repeats, respectively (Figure 3 and Tables S9–S12). For the majority of the repeats, their sizes are: In *C. farinosa*—20–29 bp (69.38%), followed by 10–19 bp (22.44%), followed by 30–39 bp (4.08%), whereas 40–49 bp and

60–69 bp are the least common, at 2.04%. In *C. glandulosa*—20–29 bp (87.75%), followed by 30–39 bp (6.12%), whereas 10–19 bp, 40–49 bp and 60–69 bp are the least common, at 2.04%. In *M. crassifolia*—20–29 bp (48.97%), followed by 10–19 bp (38.77%), with 50–59 bp and 40–49 bp being the least common, at 6.12% and 4.08%, respectively, whereas 30–39 bp was at 2.04%. In *M. oblongifolia*—20–29 bp (65.30%), followed by 10–19 bp (26.53%), followed by 50–59 bp (4.08%), whereas 30–39 bp and 40–49 bp are the least common, at 2.04%. In total, there are 49 repeats in the chloroplast genomes of the four species. In the first location, the codon region harbored 42.85% of the repeats in *C. farinosa*, *M. crassifolia* and *M. oblongifolia* and 34.69% in *C. glandulosa*; tRNA contained 7 repeats (14.28%) in *C. farinosa*, 8 repeats (16.32%) in *C. glandulosa*, 9 repeats (18.36%) in *M. crassifolia* and 10 repeats (20.40%) in *M. oblongifolia*; the remainder of the repeats are located in the protein-coding genes—7 repeats (14.28%) in *C. farinosa* and *C. glandulosa*, 6 repeats (12.24%) in *M. crassifolia* and 12 repeats (24.48%) in *M. oblongifolia*. The length of repeated sequences in the four Capparaceae chloroplast genomes ranged from 10 to 59 bp, analogous to the lengths in other angiosperm plants [62–64].

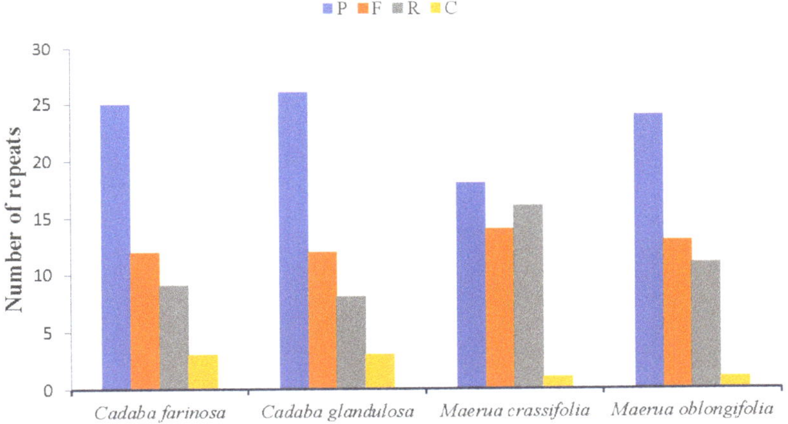

Figure 3. Number of different repeats in four chloroplast genomes of four species of Capparaceae. *p* = palindromic, F = forward, R = reverse and C= complement.

2.4.2. Simple Sequence Repeats (SSRs)

The SSRs or microsatellites are a group of short repeat sequences of nucleotide series (1–6 bp), which are used as a tool to facilitate the assessment of molecular diversity [65]. The genetic variation within and among species with the valuable molecular marker of the SSRs is extremely important for studying genetic heterogeneity and contributes to species recognition [66–68]. In this study, there are 249 microsatellites found in the plastid genome of *C. farinosa*, in *C. glandulosa* there are 251, in *M. crassifolia* there are 227 and in *M. oblongifolia* there are 233 (Table 3). The majority of SSRs in the cp genome in *C. farinosa*, *C. glandulosa*, *M. crassifolia* and *M. oblongifolia* are mononucleotide (88.75%, 89.24%, 90.74% and 90.12%, respectively), of which most are poly T and A (Figure 4). Polythymine (poly T) constituted 50.60%, 52.19%, 51.98% and 52.78%, respectively, whereas polyadenine (poly A) constituted 37.75%, 36.65%, 37.88% and 36.48%, respectively. Only a single polycytosine (poly C) (0.40% and 0.42%) was present in *C. farinosa* and *M. oblongifolia*, whereas two (0.88%) were present in *M. crassifolia*, and only a single polyguanine (poly G) (0.39% and 0.42%) was present in *C. glandulosa* and *M. oblongifolia*. Among the dinucleotides, AT/AT, AC/GT and AG/CT were found in all genomes. Reflecting series complementary, only one trinucleotide, AAT/ATT, six tetranucleotides, AAAC/GTTT, AAAG/CTTT, AAAT/ATTT, AATT/AATT, AACT/AGTT and AGAT/ATCT, and five pentanucleotides, AAAAT/ATTTT, AAATT/AATTT, AACAT/ATGTT, AAACT/AGTTT and AATAG/ATTCT, were discovered in the genome, while no hexanucleotide repeat was present (Figure 4). A high richness in mononucleotides poly A and T has been observed in most flowering plants' cp genomes [62].

Table 3. Simple sequence repeats in the *C. farinosa*, *C. glandulosa*, *M. crassifolia* and *M. oblongifolia* chloroplast genomes.

SSR Type	Repeat Unit	Species			
		C. farinosa	*C. glandulosa*	*M. crassifolia*	*M. oblongifolia*
Mono	A/T	220	223	204	208
	C/G	1	1	2	2
Di	AC/GT	0	1	0	0
	AG/CT	2	0	1	1
	AT/AT	11	12	7	10
Tri	AAT/ATT	4	5	2	3
Tetra	AAAC/GTTT	0	0	1	0
	AAAG/CTTT	0	0	0	1
	AAAT/ATTT	6	6	4	5
	AATT/AATT	2	2	1	0
	AACT/AGTT	0	0	0	1
	AGAT/ATCT	1	1	2	1
Penta	AAAAT/ATTTT	1	0	0	0
	AAATT/AATTT	0	0	1	0
	AACAT/ATGTT	0	0	0	1
	AAACT/AGTTT	1	0	0	0
	AATAG/ATTCT	0	0	2	0

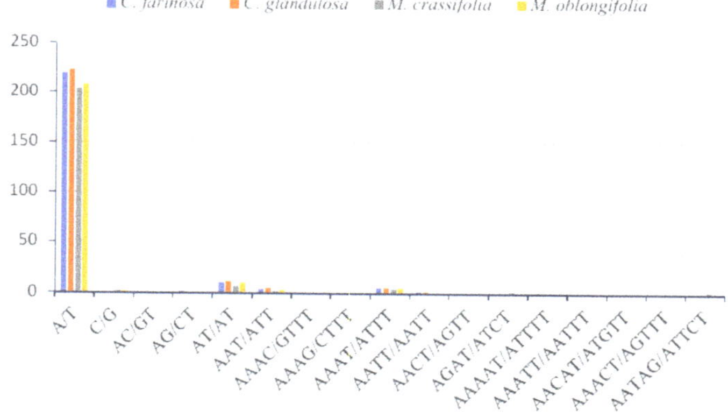

Figure 4. Frequency of different SSR motifs in different repeat types in *C. farinosa*, *C. glandulosa*, *M. crassifolia* and *M. oblongifolia* chloroplast genomes.

The comparison of simple sequence repeats between the chloroplast genomes of the four Capparaceae species (Figure 5) indicated that the more frequent occurrences are the mononucleotide repeats in all the genomes. The largest number of mononucleotides in

C. glandulosa was 224, while it did not possess a pentanucleotide, like the remaining three species. Hexanucleotide was not present in any of the four species.

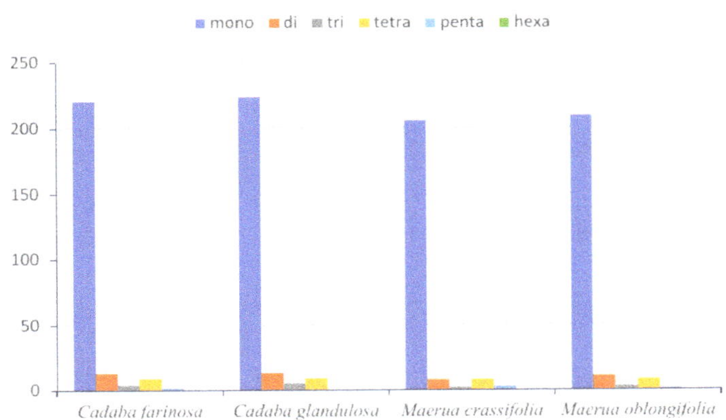

Figure 5. Number of different SSR types in the four chloroplast genomes of Capparaceae.

2.5. Comparative Analysis of the Capparaceae Species Cp Genome

To analyze the DNA sequence divergence in the chloroplast genomes of the five species of Capparaceae, a comparative analysis was done using the mVISTA program to align the sequences. Sequence alignment was conducted among four chloroplast genomes of Capparaceae and compared with the chloroplast genome of *Capparis versicolor* (MH142726), available in GenBank. To understand the structural characteristics in the cp genomes, the annotation of *C. farinosa* was used as a reference. The alignment outcome reveals highly conserved genomes with few variations. As in most chloroplast genomes of angiosperm plants, non-coding counterparts were conserved less than the gene-coding regions (Figure 6). Among the five cp genomes, the results showed that *trnH(GUG)-psbA, rps16-trnQ, psbI-trnS, trnS-trnR, petN-psbM, psbM-trnD, trnE-trnT, trnS-trnG, trnT-trnL, trnF-ndhJ, rbcL-accD, psbE-petL, rbs16-rbs3* and *ndhF-rpl32* were the most divergent non-coding regions. However, it was detected that some variations occurred in the following genes: *atpF, rpoC2, rps19* and *ycf1*.

Although angiosperms retain the structure and size of the chloroplast genome [68], some evolutionary events occur in the genome, such as expansion and contraction, that alter the size of the genome and the boundaries of the LSC, SSC, IRa and IRb regions [69,70]. We compared between IR-LCS and IR-SSC the boundaries of the five cp genomes of Capparaceae (*Cadaba farinosa, Cadaba glandulosa, Maerua crassifolia, Maerua oblongifolia* and *Capparis versicolor*) and the result presented a similarity among the compared plastomes of *Cadaba* and *Maerua* species, with a slight variation among *C. versicolor* (Figure 7). The chloroplast genome of *C. versicolor* (155,051 bp) was the smallest, whereas the genome of *C. glandulosa* (156,560 bp) was the largest. The smallest IR region is in *C. versicolor* (26,141 bp). The lengths of LSC regions varied among the five Capparaceae species (85,565 bp, 85,681 bp, 84,624 bp, 84,153 bp, 84,315 bp, respectively). The location of the *rpsl9* gene is between the junction of the LSC and IRb regions in five species and is in the LSC region in *C. versicolor*. The *ycf1* gene is located in IRb regions, except in *C. versicolor*, and it crosses the SSC/IRa region and extends by different lengths into the SSC region within the genome (*C. farinosa* and *C. glandulosa* 4360 bp; *M. crassifolia* 4393 bp; *M. oblongifolia* 4414 bp and *C. versicolor* 4566 bp). The *ndhF* gene is found in the IRb/SSC and is 38 bp in *C. farinosa* and *C. glandulosa*, 32 bp in *M. crassifolia* and 35 bp in *M. oblongifolia* in the IRb region, and it extends into the SSC region by 2209 bp in *C. farinosa* and *C. glandulosa* and 2206 bp in *M. crassifolia* and *M. oblongifolia*, and is 174 bp away from the border in the *C. versicolor* genome.

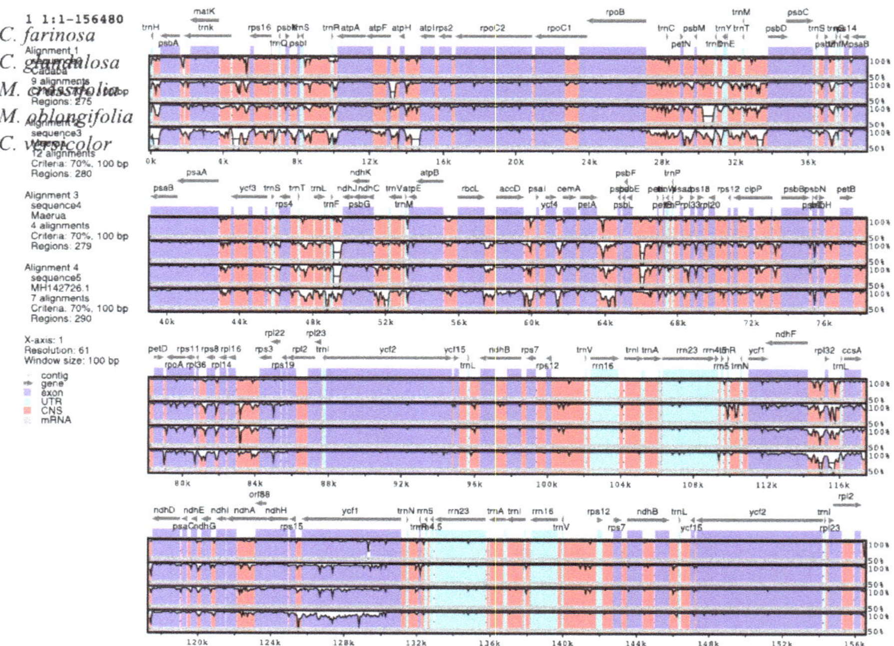

Figure 6. Alignment of chloroplast genomes of *C. farinosa*, *C. glandulosa*, *M. crassifolia*, *M. oblongifolia* and *C. versicolor* performed with *C. farinosa* as reference. Transcription direction is indicated by the gray arrows at the top, protein coding is represented by blue bars, non-coding sequence CNS is represented by pink bars and tRNAs and rRNAs are represented by light green. The cp genome is identified by the coordinates in the x-axis, while the y-axis represents the percentage identity within 50–100%.

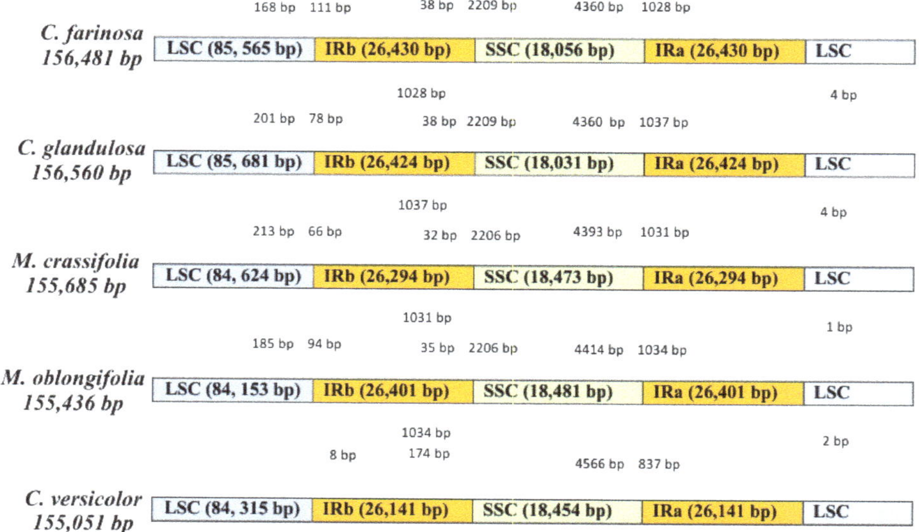

Figure 7. Comparison of the IR, SSC and LSC junction positions among five chloroplast genomes of Capparaceae.

2.6. Divergence of Protein-Coding Gene Sequence

The cp genomes of four Capparaceae species include 80 protein-coding genes in *C. glandulosa* and 81 genes in other species. To detect the genes under selective pressure, the rates of synonymous (dS) and non-synonymous (dN) substitution and dN/dS ratio were calculated. The results showed that in all of the paired genes of *C. farinosa* vs. *C. glandulosa*, the dN/dS ratio is less than 1, and most of the paired genes are less than 1 except *atpF* in *C. farinosa* vs. *M. crassifolia* and *cemA, psbK* and *rps18* in *C. farinosa* vs. *M. oblongifolia*, having values of 1.16, 1.52 and 1.2, respectively (Figure 8). The result of the dN/dS ratio obtained in this study is consistent with other related studies [52,53]. In all the genes, the synonymous (dS) values range from 0 to 0.32 (Figure 8).

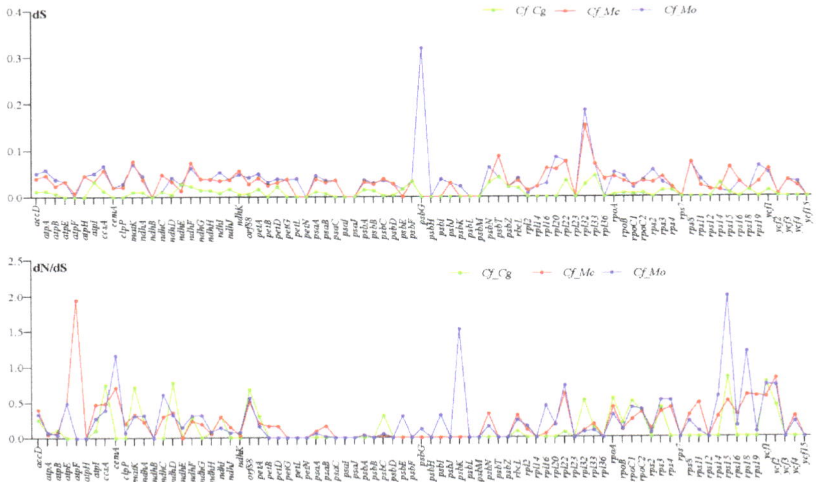

Figure 8. The synonymous (dS) and dN/dS ratio values of 81 protein-coding genes from four Capparaceae cp genomes.

2.7. Phylogenetic Analysis

Phylogenetic relationships based on Bayesian analysis and maximum parsimony were congruent and placed all samples into three main clades, with strong support in all the nodes with PP 1.00 (Figure 9). The first clade contains species of the Capparaceae family and is divided into two subclades; the first subclade includes species of genera *Cadaba* and *Maerua*, while the second clade includes species of genus *Capparis*. The second clade comprises Cleomaceae species, while the third clade includes species from the Brassicaceae family. The phylogenetic tree showed that the Capparaceae family is the earliest diverging lineage among the three families and is sister to Cleomaceae and Brassicaceae. It is clear in this phylogenetic result that Cleomaceae was separated from Capparaceae and became a sister to the Brassicaceae family, as reported by [19,20]; this is consistent with some previous classifications of the order Brassicales.

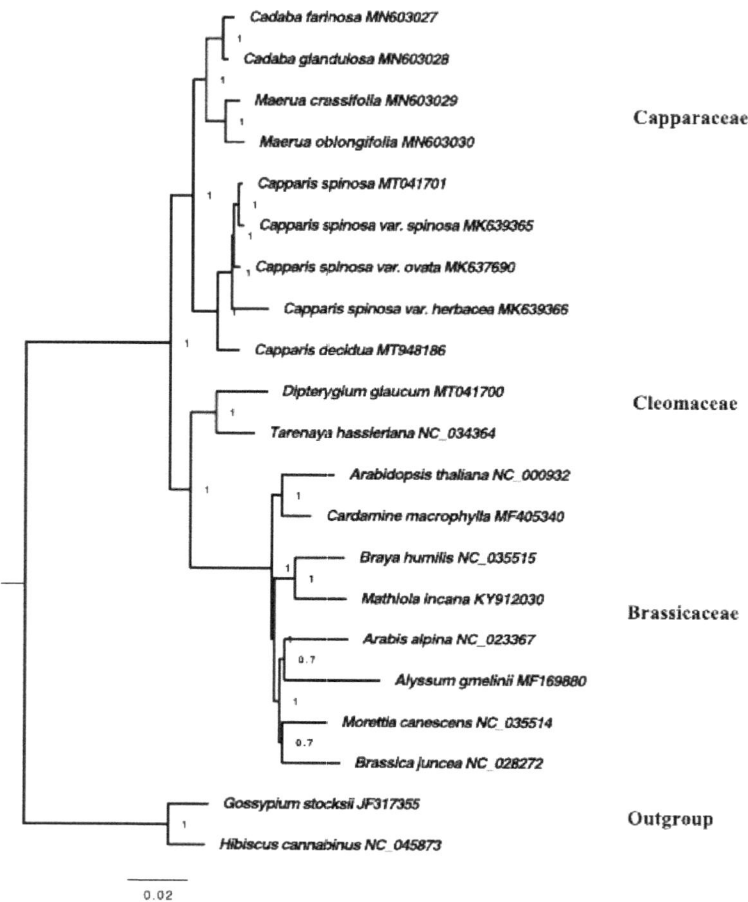

Figure 9. Phylogenetic tree reconstruction based on the complete chloroplast genome of 21 taxa inferred from Bayesian inference (BI) methods, showing relationships within Brassicales. Numbers in the clade represent posterior probability (PP) values.

3. Materials and Methods

3.1. Plant Material and DNA Extraction

Fresh young leaves were collected in 2018 during field investigations in Saudi Arabia: *C. farinosa* in Jeddah (21°26′45.2″ N 39°25′22.9″ E) on 21 April, *C. glandulosa* in Jeddah (21°26′45.3″ N 39°25′22.9″ E) on 21 April, *M. crassifolia* in Makkah (21°13′17.4″ N 39°49′36.1″ E) on 28 April and *M. oblongifolia* in Jeddah (21°28′31.7″ N 39°50′36.8″ E) on 21 April. No permission was required to collect the plant samples. Species were identified and verified by Dr. Dhafer Alzahrani, Department of Biological Sciences, Faculty of Sciences, King Abdulaziz University, Jeddah, Saudi Arabia. A voucher specimen was prepared and deposited in the herbarium of King Abdulaziz University, Jeddah with the accession numbers: *C. farinosa* (KAU27480), *C. glandulosa* (KAU27481) *M. crassifolia* (KAU27482), *M. oblongifolia* (KAU27483). Total genomic DNA was extracted from the samples using the Qiagen genomic DNA extraction kit, according to the manufacturer's protocols.

3.2. Library Construction, Sequencing and Assembly

Input material for the DNA sample preparations was derived (or taken) from a total amount of 1.0 µg DNA. The NEBNext DNA Library Prep Kit was used to generate sequence libraries according to the manufacturer's recommendation; indices were also added to each sample. Genomic DNA was randomly fragmented by shearing to a size of 350 bases in length. The ends of randomly fragmented DNA were repaired and A-tailed, adapters were ligated with NEBNext for Illumina sequencing, then the PCR improved by P5 and indexed P7 oligo sequences. The AMPure XP system was used to purify the PCR products; subsequent findings were analyzed by the Agilent 2100 Bioanalyzer for size distribution and later quantified using real-time PCR. After pooling, the qualified libraries were fed into an Illumina HiSeq 2500 system (350 bp paired ends reads); this was based on its effective concentration and expected data volume. The raw reads (19,844,190 bp, 19,053,503 bp, 19,440,639 bp and 19,929,468 bp for *C. farinosa*, *C. glandulosa*, *M. crassifolia* and *M. oblongifolia*, respectively) were cleaned reads (5 Gb) to remove low-quality sequences and adaptors; they were then filtered for PCR duplicates using PRINSEQlite v0.20.4 [71]. The clean raw reads were subjected to de novo assembly from the whole genome sequences using NOVOPlasty 2.7.2 [72] with kmer (K-mer = 31–33). The *trnH-psbA* of *Cadaba farinosa* (KR735837.1) was used as a seed and the complete plastome of *Arabidopsis thaliana* (KX551970.1) was used as a reference for the assembly of the *Cadaba farinosa* cp genome. The assembled cp genome of *Cadaba farinosa* was used as seed and reference for the assembly of the *Cadaba glandulosa* plastome. For *Maerua crassifolia*, the *rpoC1* gene of *M. crassifolia* (JQ845894.1) was used as seed and the complete cp genome of *C. farinosa* was used as reference. The assembled cp genome of *M. crassifolia* was used as seed and reference for the assembly of *M. oblongifolia*. Finally, each species generated one contig that contained the complete chloroplast genome sequence.

3.3. Gene Annotation

Genes were annotated using the Dual Organellar GenoMe Annotator (DOGMA, University of Texas at Austin, Austin, TX, USA) [73]. The positions of start and stop codons were adjusted manually. tRNA genes were identified by the trnAscan-SE server (http://lowelab.ucsc.edu/tRNAscan-SE/ (accessed on 20 June 2019) [74]. Organellar Genome DRAW (OGDRAW) [75] was used to draw the genome maps.

3.4. Sequence Analysis

MEGA 6.0 was used to compute the codon usage, base composition, and the relative synonymous codon usage values (RSCUs). The RNA editing sites in cp protein-coding genes of the Capparaceae species were predicted using PREP suite [76] with a 0.8 cutoff value. Simple sequence repeats (SSRs) were identified in the chloroplast genomes of the four species (*C. farinosa*, *C. glandulosa*, *M. crassifolia* and *M. oblongifolia*) using the online software MIcroSAtellite (MISA) [77] with the following parameters set: eight, five, four and three repeat units for mononucleotides, dinucleotides, trinucleotides and tetra-, penta-, hexanucleotide SSR motifs, respectively. To identify the size and location of long repeats (palindromic, forward, reverse and complement) in the four species of Capparaceae being studied, the online program REPuter (https://bibiserv.cebitec.uni-bielefeld.de/reputer (accessed on 22 June 2019) [76], with standard settings, was used.

3.5. Genome Comparison

The chloroplast genomes of *C. farinosa*, *C. glandulosa*, *M. crassifolia* and *M. oblongifolia* were compared using the program mVISTA [78], and the annotation of *C. farinosa* was used as a reference in the Shuffle-LAGAN mode [79]. The four species of Capparaceae were compared against the border region between inverted repeat (IR), large single copy (LSC) and small single copy (SSC).

3.6. Characterization of Substitution Rate

To detect the genes that are under selection pressure, the substitution rate of the synonymous (dS) and non-synonymous (dN) substitution and the dN/dS ratio were analyzed using DNAsp v5.10.01 [30], the cp genome of *C. farinosa* was compared with the cp genome of *C. glandulosa*, *M. crassifolia* and *M. oblongifolia*. Separate protein-coding genes were aligned individually using Geneious version 8.1.3 software, while the protein sequence was translated from aligned sequences.

3.7. Phylogenetic Analysis

The analysis was conducted based on the complete chloroplast genome sequences of nine Capparaceae taxa, six species and three varieties, *C. farinosa*, MN603027, *C. glandulosa*, MN603028, *M. crassifolia*, MN603029, *M. oblongifolia*, MN603030, *Capparis spinosa*, MT041701, *Capparis spinosa* var. *spinosa*, MK639365, *Capparis spinosa* var. *herbacea*, MK639366, *Capparis spinosa* var. *ovata*, MK637690 and *Capparis decidua* MT948186, two Cloemaceae species, eight species of Brassicaceae and two species of Malvaceae, as an outgroup. All sequences were aligned using MAFFT software [81] with default settings. The phylogenetic trees were reconstructed based on maximum parsimony analysis using PAUP software (version 4.0b10) [82], utilizing tree bisection and reconnection branch swapping, with MulTrees on, saving a maximum of 1000 trees per replicate. Missing characters were treated as gaps. The bootstrap analysis confidence was based on 1000 replicates. MrBayes version 3.2.6 [83] was used to conduct Bayesian analysis and jModelTest version 3.7 [84] was used to select the appropriate model.

4. Conclusions

This current study used the Illumina HiSeq 2500 platform to obtain the first complete chloroplast sequences of four medicinal Capparaceae species: *C. farinosa*, *C. glandulosa*, *M. crassifolia* and *M. oblongifolia*. The four species are divided into two groups: *C. farinosa* and *C. glandulosa* belong to the tribe Cadabeae; *M. crassifolia* and *M. oblongifolia* belong to the tribe Maerueae. This study can be used to accurately identify species during different medicinal uses based on their plastid genome.

Supplementary Materials: The following are available online at https://www.mdpi.com/article/10.3390/plants10061229/s1, Table S1: Codon–anticodon recognition patterns and codon usage of the *Cadaba farinosa* chloroplast genome; Table S2: Codon–anticodon recognition patterns and codon usage of the *Cadaba glandulosa* chloroplast genome; Table S3: Codon–anticodon recognition patterns and codon usage of the *Maerua crassifolia* chloroplast genome; Table S4: Codon–anticodon recognition patterns and codon usage of the *Maerua oblongifolia* chloroplast genome; Table S5: Predicted RNA editing site in the *Cadaba farinosa* chloroplast genome; Table S6: Predicted RNA editing site in the *Cadaba glandulosa* chloroplast genome; Table S7: Predicted RNA editing site in the *Maerua crassifolia* chloroplast genome; Table S8: Predicted RNA editing site in the *Maerua oblongifolia* chloroplast genome; Table S9: Repeat sequences present in the *Cadaba farinosa* chloroplast genome; Table S10: Repeat sequences present in the *Cadaba glandulosa* chloroplast genome; Table S11: Repeat sequences present in the *Maerua crassifolia* chloroplast genome; Table S12: Repeat sequences present in the *Maerua oblongifolia* chloroplast genome.

Author Contributions: D.A.A. and E.J.A. designed the research and performed the experiments, D.A.A., S.S.Y. and A.A. collected data, E.J.A. and S.S.Y. analyzed the data and drafted the manuscript, D.A.A. supervised the project. All authors have read and agreed to the published version of the manuscript.

Funding: This project was funded by the Deanship of Scientific Research (DSR), King Abdulaziz University, Jeddah, under grant no. DF-294-130-1441. The authors, therefore, gratefully acknowledge DSR technical and financial support.

Data Availability Statement: The complete chloroplast genome sequence of four Capparaceae chloroplast genome sequences were deposited in GenBank at https://www.ncbi.nlm.nih.gov, (accession numbers: *C. farinosa*, MN603027; *C. glandulosa*, MN603028; *M. crassifolia*, MN603029 and *M. oblongifolia*, MN603030).

Conflicts of Interest: The authors declare no conflict of interest.

References

1. Cronquist, A.; Takhtadzhián, A.L. *An Integrated System of Classification of Flowering Plants*; Columbia University Press: New York, NY, USA, 1981; p. 337.
2. Mabberley, D.J. *The Plant-Book: A Portable Dictionary of the Higher Plants*; Cambridge University Press: Cambridge, UK, 1997.
3. Heywood, V.H.; Brummitt, R.K.; Culham, A.; Seberg, O. Flowering Plant Families of the World. Royal Botanic Gardens, Kew. *Int. J. Pharm.* **2007**, *11*, 874–887.
4. Gibbs, R.D. *Chemotaxonomy of Flowering Plants*; McGill-Queens University Press: London, UK, 1974; p. 761.
5. Rahman, M.; Mossa, J.S.; Al-Said, M.S.; Al-Yahya, M.A. Medicinal plant diversity in the flora of Saudi Arabia 1: A report on seven plant families. *Fitoterapia* **2004**, *75*, 149–161. [CrossRef]
6. Telrandhe, U.B.; Uplanchiwar, V. Phyto-Pharmacological Perspective of Cadaba farinosa forsk. *Am. J. Phytomed. Clin. Ther.* **2013**, *1*, 11–22.
7. Agize, M.; Demissew, S.; Asfaw, Z. Ethnobotany of Medicinal Plants in Loma and Gena Bosa Districts (Woredas) of Dawro Zone, Southern Ethiopia. *Topclass J. Herb. Med.* **2013**, *2*, 194–212.
8. Laxmichand, B.H.; Modi, N.R. A Comprehensive Review on Maerua Oblongifolia (Forsk.) A. Rich. *Int. J. Res. Advent Technol.* **2019**, *4*, 2321–9637.
9. Burkill, H.M.; Dalziel, J.M.; Hutchinson, J. Entry for *Maerua crassifolia* Forssk. [family CAPPARACEAE]. In *Useful Plants of West Tropical Africa*, 2nd ed.; Royal Botanic Gardens: Kew, UK, 1985.
10. Akuodor, G.C.; Ibrahim, J.A.; Akpan, J.L.; Okorie, A.U.; Ezeokpo, B.C. Phytochemical and Anti-diarrhoeal Properties of Methanolic Leaf Extract of *Maerua crassifolia* Forssk. *Eur. J. Med. Plants* **2014**, *4*, 1223–1231. [CrossRef]
11. Ckilaka, K.; Akuodor, G.; Akpan, J.; Ogiji, E.; Eze, C.; Ezeokpo, B.C. Antibacterial and antioxidant activities of methanolic leaf extract of *Maerua crassifolia*. *J. Appl. Pharm. Sci.* **2015**, *5*, 147–150. [CrossRef]
12. Bentham G.; Hooker, J.D. *Genera Plantarum ad Eemplaria Imprimis in Herbariis Kewensibus Servata Definite*; Reeve and Co.: London, UK, 1883.
13. Engler, A.; Prantl, K. *Die Naturlichen Pflanzenfamilien*; Wilhelm Engelmann: Leipzig, Germany, 1915.
14. Hutchinson, J. *The Genera of Flowering Plants*; Clarendon Press: Oxford, UK, 1967.
15. Takhtajan, A. *Origins of Angiospermous Plants*; American Institute of Biological Science: Washington, DC, USA, 1954.
16. Nee, M. Diversity and Classification of Flowering Plants A. Takhtajan. *Brittonia* **1998**, *50*, 191–192. [CrossRef]
17. APG I (Angiosperm Phylogeny Group). An ordinal classification for the families of flowering plants. *Ann. Mo. Bot. Gard.* **1998**, *85*, 531–553. [CrossRef]
18. APG II (Angiosperm Phylogeny Group). An update of the Angiosperm Phylogeny Group classification for the orders and families of flowering plants. *Bot. J. Linn. Soc.* **2003**, *141*, 399–436. [CrossRef]
19. APG III (Angiosperm Phylogeny Group). An update of the Angiosperm Phylogeny Group classification for the orders and families of flowering plants. *Bot. J. Linn. Soc.* **2009**, *161*, 105–121.
20. APG IV (Angiosperm Phylogeny Group). An update of the Angiosperm Phylogeny Group classification for the orders and families of flowering plants. *Bot. J. Linn. Soc.* **2016**, *181*, 1–20.
21. Iltis, H.E. Studies in the Capparidaceae. III. Evolution and Phylogeny of the Western North American Cleomoideae. *Ann. Mo. Bot. Gard.* **1957**, *44*, 77. [CrossRef]
22. Al-Shehbaz, I.A. The biosystematics of the genus Thelypodium (Cruciferae). *Contrib. Gray Herb. Harv. Univ.* **1973**, *204*, 3–148.
23. Al-Shehbaz, I.A. The tribes of Cruciferae (Brassicaceae) in the southeastern United States. *J. Arn. Arbor.* **1984**, *65*, 343–373.
24. Dalhgren, R. A System of classification of the angiosperms to be used to demonstrate the distribution of characters. *Bot. Not.* **1975**, *128*, 119–147.
25. Takhtajan, A.L. Outline of the classification of flowering plants (magnoliophyta). *Bot. Rev.* **1980**, *46*, 225–359. [CrossRef]
26. Hauser, L.A.; Crovello, T.J. Numerical Analysis of Generic Relationships in Thelypodieae (Brassicaceae). *Syst. Bot.* **1982**, *7*, 249. [CrossRef]
27. Rodman, J.; Price, R.A.; Karol, K.; Conti, E.; Systma, K.J.; Palmer, J.D. Nucleotide Sequences of the rbcL Gene Indicate Monophyly of Mustard Oil Plants. *Ann. Mo. Bot. Gard.* **1993**, *80*, 686. [CrossRef]
28. Rodman, J.E.; Karol, K.G.; Price, R.A.; Sytsma, K.J. Molecules, Morphology, and Dahlgren's Expanded Order Capparales. *Syst. Bot.* **1996**, *21*, 289. [CrossRef]
29. Rodman, J.E.; Soltis, P.S.; Soltis, D.E.; Sytsma, K.J.; Karol, K.G. Parallel evolution of glucosinolate biosynthesis inferred from congruent nuclear and plastid gene phylogenies. *Am. J. Bot.* **1998**, *85*, 997–1006. [CrossRef]
30. Schmid, R.; Rollins, R.C. The Cruciferae of Continental North America: Systematics of the Mustard Family from the Arctic to Panama. *TAXON* **1994**, *43*, 153. [CrossRef]

31. Judd, W.S.; Sanders, R.W.; Donoghue, M.J. Angiosperm family pairs: Preliminary phylogenetic analyses. *Harv. Pap. Bot.* **1994**, *5*, 1–51.

32. Judd, W.S.; Campbell, C.S.; Kellogg, E.A.; Stevens, P.F.; Donoghue, M.J. *Plant Systematics: A Phylogenetic Approach*, 3rd ed.; Sinauer Associates: Sunderland, MA, USA, 2007.

33. Hall, J.; Sytsma, K.J.; Iltis, H.H. Phylogeny of Capparaceae and Brassicaceae based on chloroplast sequence data. *Am. J. Bot.* **2002**, *89*, 1826–1842. [CrossRef] [PubMed]

34. Airy Shaw, H.K. Diagnoses of new families, new names, etc., for the seventh edition of Willis's 'Dictionary'. *Kew Bull.* **1965**, *18*, 249–273. [CrossRef]

35. Pax, F.; Hoffmann, K. Capparidaceae. In *Die Natu rlichen Pflanzenfamilien*, 2nd ed.; Engler, A., Prantl, K., Eds.; Wilhelm Engelmann: Leipzig, Germany, 1936; Volume 17, pp. 146–223.

36. Hedge, I.C.; Kjaer, A.; Malver, O. Dipterygium—Cruciferaeor Capparaceae? *R. Bot. Gard. Edinb.* **1980**, *38*, 247–250.

37. Doweld, A.; Reveal, J.L. New suprageneric names for vascular plants. *Phytologia* **2008**, *90*, 416–417.

38. Grevich, J.J.; Daniell, H. Chloroplast Genetic Engineering: Recent Advances and Future Perspectives. *Crit. Rev. Plant Sci.* **2005**, *24*, 83–107. [CrossRef]

39. Neuhaus, H.; Emes, M. Nonphotosynthetic metabolism in plastids. *Ann. Rev. Plant Biol.* **2000**, *51*, 111–140. [CrossRef] [PubMed]

40. WickeGerald, S.; Schneeweiss, G.M.; Depamphilis, C.W.; Müller, K.F.; Quandt, D. The evolution of the plastid chromosome in land plants: Gene content, gene order, gene function. *Plant Mol. Biol.* **2011**, *76*, 273–297. [CrossRef]

41. Raubeson, L.; Jansen, R.; Henry, R.J. (Eds.) Chloroplast genomes of plants. In *Plant Diversity and Evolution: Genotypic and Phenotypic Variation in Higher Plants*; CABI: London, UK, 2005; pp. 45–68.

42. Greiner, S.; Sobanski, J.; Bock, R. Why are most organelle genomes transmitted maternally? *BioEssays* **2015**, *37*, 80–94. [CrossRef] [PubMed]

43. GuisingerTimothy, M.M.; Chumley, T.W.; Kuehl, J.V.; Boore, J.L.; Jansen, R.K. Implications of the Plastid Genome Sequence of Typha (Typhaceae, Poales) for Understanding Genome Evolution in Poaceae. *J. Mol. Evol.* **2010**, *70*, 149–166. [CrossRef]

44. Yang, J.-B.; Tang, M.; Li, H.-T.; Zhang, Z.-R.; Li, D.-Z. Complete chloroplast genome of the genus Cymbidium: Lights into the species identification, phylogenetic implications and population genetic analyses. *BMC Evol. Biol.* **2013**, *13*, 84. [CrossRef] [PubMed]

45. Alzahrani, D.; Albokhari, E.; Yaradua, S.; Abba, A. The complete plastome sequence for the medicinal species *Capparis spinosa* L. (Capparaceae). *Gene Rep.* **2021**, *23*, 101059. [CrossRef]

46. Maurya, S.; Darshetkar, A.M.; Datar, M.N.; Tamhankar, S.; Li, P.; Choudhary, R.K. Plastome data provide insights into intra and interspecific diversity and ndh gene loss in Capparis (Capparaceae). *Phytotaxa* **2020**, *432*, 206–220. [CrossRef]

47. Chen, H.; Shao, J.; Zhang, H.; Jiang, M.; Huang, L.; Zhang, Z.; Yang, D.; He, M.; Ronaghi, M.; Luo, X.; et al. Sequencing and Analysis of Strobilanthes cusia (Nees) Kuntze Chloroplast Genome Revealed the Rare Simultaneous Contraction and Expansion of the Inverted Repeat Region in Angiosperm. *Front. Plant Sci.* **2018**, *9*, 324. [CrossRef]

48. Chang, C.C.; Lin, H.C.; Lin, I.P.; Chow, T.Y.; Chen, H.H.; Chen, W.H.; Cheng, C.H.; Lin, C.U.; Liu, S.M.; Chang, C.C.; et al. The chloroplast genome of *Phalaenopsis aphrodite* (Orchidaceae): Comparative analysis of evolutionary rate with that of grasses and its phylogenetic implications. *Mol. Biol. Evol.* **2006**, *23*, 279–291. [CrossRef]

49. Raman, G.; Park, S. The complete chloroplast genome sequence of Ampelopsis: Gene organization, comparative analysis, and phylogenetic relationships to other angiosperms. *Front. Plant Sci.* **2016**, *341*, 7. [CrossRef]

50. Park, I.; Kim, W.J.; Yeo, S.-M.; Choi, G.; Kang, Y.-M.; Piao, R.; Moon, B.C. The Complete Chloroplast Genome Sequences of *Fritillaria ussuriensis* Maxim. And *Fritillaria cirrhosa* D. Don, and Comparative Analysis with Other *Fritillaria* Species. *Molecules* **2017**, *22*, 982. [CrossRef]

51. Li, B.; Lin, F.; Huang, P.; Guo, W.; Zheng, Y. Complete Chloroplast Genome Sequence of Decaisnea insignis: Genome Organization, Genomic Resources and Comparative Analysis. *Sci. Rep.* **2017**, *7*, 10073. [CrossRef]

52. Liu, X.; Li, Y.; Yang, H.; Zhou, B. Chloroplast Genome of the Folk Medicine and Vegetable Plant Talinum paniculatum (Jacq.) Gaertn.: Gene Organization, Comparative and Phylogenetic Analysis. *Molecules* **2018**, *23*, 857. [CrossRef]

53. Mower, J.P. The PREP suite: Predictive RNA editors for plant mitochondrial genes, chloroplast genes and user-defined aligrments. *Nucleic Acids Res.* **2009**, *37*, W253–W259. [CrossRef] [PubMed]

54. Bundschuh, R.; Altmüller, J.; Becker, C.; Nürnberg, P.; Gott, J.M. Complete characterization of the edited transcriptome of the mitochondrion of Physarum polycephalum using deep sequencing of RNA. *Nucleic Acids Res.* **2011**, *39*, 6044–6055. [CrossRef]

55. Zeng, W.-H.; Liao, S.-C.; Chang, C.-C. Identification of RNA Editing Sites in Chloroplast Transcripts of Phalaenopsis aphrodite and Comparative Analysis with Those of Other Seed Plants. *Plant Cell Physiol.* **2007**, *48*, 362–368. [CrossRef] [PubMed]

56. Luo, J.; Hou, B.-W.; Niu, Z.-T.; Liu, W.; Xue, Q.-Y.; Ding, X.-Y. Comparative Chloroplast Genomes of Photosynthetic Orchids: Insights into Evolution of the Orchidaceae and Development of Molecular Markers for Phylogenetic Applications. *PLoS ONE* **2014**, *9*, e99016. [CrossRef] [PubMed]

57. Wang, W.; Yu, H.; Wang, J.; Lei, W.; Gao, J.; Qiu, X.; Wang, J. The Complete Chloroplast Genome Sequences of the Medicinal Plant *Forsythia suspensa* (Oleaceae). *Int. J. Mol. Sci.* **2017**, *18*, 2288. [CrossRef]

58. Kumbhar, F.; Nie, X.; Xing, G.; Zhao, X.; Lin, Y.; Wang, S.; Weining, S. Identification and characterisation of RNA editing sites in chloroplast transcripts of einkorn wheat (Triticum monococcum). *Ann. Appl. Biol.* **2018**, *172*, 197–207. [CrossRef]

59. Park, M.; Park, H.; Lee, H.; Lee, B.-H.; Lee, J. The Complete Plastome Sequence of an Antarctic Bryophyte *Sanionia uncinata* (Hedw.) Loeske. *Int. J. Mol. Sci.* **2018**, *19*, 709. [CrossRef]
60. Magdalena, G.N.; Ewa, F.; Wojciech, P. Cucumber, melon, pumpkin, and squash: Are rules of editing in flowering plants chloroplast genes so well-known indeed? *Gene* **2009**, *434*, 1–8.
61. Huang, Y.-Y.; Matzke, A.J.M.; Matzke, M. Complete Sequence and Comparative Analysis of the Chloroplast Genome of Coconut Palm (Cocos nucifera). *PLoS ONE* **2013**, *8*, e74736. [CrossRef]
62. Li, Y.; Xu, W.; Zou, W.; Jiang, D.; Liu, X. Complete chloroplast genome sequences of two endangered Phoebe (Lauraceae) species. *Bot. Stud.* **2017**, *58*, 1–10. [CrossRef]
63. Greiner, S.; Wang, X.; Rauwolf, U.; Silber, M.V.; Mayer, K.; Meurer, J.; Haberer, G.; Herrmann, R.G. The complete nucleotide sequences of the five genetically distinct plastid genomes of Oenothera, subsection Oenothera: I. Sequence evaluation and plastome evolution. *Nucleic Acids Res.* **2008**, *36*, 2366–2378. [CrossRef] [PubMed]
64. Song, Y.; Wang, S.; Ding, Y.; Xu, J.; Li, M.F.; Zhu, S.; Chen, N. Chloroplast Genomic Resource of Paris for Species Discrimination. *Sci. Rep.* **2017**, *7*, 3427. [CrossRef]
65. Kaila, T.; Chaduvla, P.K.; Rawal, H.C.; Saxena, S.; Tyagi, A.; Mithra, S.V.A.; Solanke, A.U.; Kalia, P.; Sharma, T.R.; Singh, N.K.; et al. Chloroplast Genome sequence of Cluster bean (Cyamopsis tetragonoloba L.): Genome structure and comparative analysis. *Genes* **2017**, *8*, 212. [CrossRef]
66. Bryan, G.J.; McNicoll, J.; Ramsay, G.; Meyer, R.C.; De Jong, W.S. Polymorphic simple sequence repeat markers in chloroplast genomes of Solanaceous plants. *Theor. Appl. Genet.* **1999**, *99*, 859–867. [CrossRef]
67. Provan, J. Novel chloroplast microsatellites reveal cytoplasmic variation in Arabidopsis thaliana. *Mol. Ecol.* **2000**, *9*, 2183–2185. [CrossRef] [PubMed]
68. Ebert, D.; Peakall, R. Chloroplast simple sequence repeats (cpSSRs): Technical resources and recommendations for expanding cpSSR discovery and applications to a wide array of plant species. *Mol. Ecol. Resour.* **2009**, *9*, 673–690. [CrossRef]
69. Philippe, H.; Delsuc, F.; Brinkmann, H.; Lartillot, N. Phylogenomics, Annual Review of Ecology. *Evol. Syst.* **2005**, *36*, 541–562. [CrossRef]
70. Raubeson, L.A.; Peery, R.; Chumley, T.W.; Dziubek, C.; Fourcade, H.M.; Boore, J.L.; Jansen, R.K. Comparative chloroplast genomics: Analyses including new sequences from the angiosperms *Nuphar advena* and *Ranunculus macranthus*. *BMC Genom.* **2007**, *8*, 1–27. [CrossRef] [PubMed]
71. Schmieder, R.; Edwards, R. Quality control and preprocessing of metagenomic datasets. *Bioinformatics* **2011**, *27*, 863–864. [CrossRef]
72. Dierckxsens, N.; Mardulyn, P.; Smits, G. NOVOPlasty: De novo assembly of organelle genomes from whole genome data. *Nucleic Acids Res.* **2017**, *45*, e18.
73. Wyman, S.K.; Jansen, R.K.; Boore, J.L. Automatic annotation of organellar genomes with DOGMA. *Bioinformatics* **2004**, *20*, 3252–3255 [CrossRef]
74. Schattner, P.; Brooks, A.N.; Lowe, T.M. The tRNAscan-SE, snoscan and snoGPS web servers for the detection of tRNAs and snoRNAs. *Nucleic Acids Res.* **2005**, *33*, W686–W689. [CrossRef] [PubMed]
75. Lohse, M.; Drechsel, O.; Bock, R. OrganellarGenomeDRAW (OGDRAW): A tool for the easy generation of high-quality custom graphical maps of plastid and mitochondrial genomes. *Curr. Genet.* **2007**, *52*, 267–274. [CrossRef] [PubMed]
76. Kurtz, S.; Choudhuri, J.V.; Ohlebusch, E.; Schleiermacher, C.; Stoye, J.; Giegerich, R. REPuter: The manifold applications of repeat analysis on a genomic scale. *Nucleic Acids Res.* **2001**, *29*, 4633–4642. [CrossRef] [PubMed]
77. Thiel, T.; Michalek, W.; Varshney, R.; Graner, A. Exploiting EST databases for the development and characterization of gene-derived SSR-markers in barley (*Hordeum vulgare* L.). *Theor. Appl. Genet.* **2003**, *106*, 411–422. [CrossRef] [PubMed]
78. Mayor, C.; Brudno, M.; Schwartz, J.R.; Poliakov, A.; Rubin, E.M.; Frazer, K.A.; Pachter, L.S.; Dubchak, I. VISTA: Visualizing global DNA sequence alignments of arbitrary length. *Bioinformatics* **2000**, *16*, 1046–1047. [CrossRef] [PubMed]
79. Frazer, K.A.; Pachter, L.; Poliakov, A.; Rubin, E.M.; Dubchak, I. VISTA: Computational tools for comparative genomics. *Nucleic Acids Res.* **2004**, *32*, W273–W279. [CrossRef]
80. Librado, P.; Rozas, J. DnaSP v5: A software for comprehensive analysis of DNA polymorphism data. *Bioinformatics* **2009**, *25*, 1451–1452. [CrossRef]
81. Katoh, K.; Standley, D.M. MAFFT multiple sequence alignment software version 7: Improvements in performance and usability. *Mol. Biol. Evol.* **2013**, *30*, 772–780. [CrossRef]
82. Felsenstein, J. Cases in which parsimony or compatibility methods will be positively misleading. *Syst. Zool.* **1978**, *27*, 401–410. [CrossRef]
83. Ronquist, F.; Teslenko, M.; van der Mark, P.; Ayres, D.L.; Darling, A.; Höhna, S.; Larget, B.; Liu, L.; Suchard, M.A.; Huelsenbeck, J.P. MrBayes 3.2: Efficient Bayesian Phylogenetic Inference and Model Choice Across a Large Model Space. *Systematic* **2012**, *61*, 539–542. [CrossRef] [PubMed]
84. Posada, D. jModelTest: Phylogenetic Model Averaging. *Mol. Biol. Evol.* **2008**, *25*, 1253–1256. [CrossRef] [PubMed]

MDPI

St. Alban-Anlage 66

4052 Basel

Switzerland

Tel. +41 61 683 77 34

Fax +41 61 302 89 18

www.mdpi.com

Plants Editorial Office

E-mail: plants@mdpi.com

www.mdpi.com/journal/plants